ABRUPT IMPACTS OF CLIMATE CHANGE

ANTICIPATING SURPRISES

Committee on Understanding and Monitoring Abrupt Climate Change and its Impacts

Board on Atmospheric Sciences and Climate

Division on Earth and Life Studies

NATIONAL RESEARCH COUNCIL
OF THE NATIONAL ACADEMIES

THE NATIONAL ACADEMIES PRESS
Washington, D.C.
www.nap.edu

THE NATIONAL ACADEMIES PRESS • 500 Fifth Street, NW • Washington, DC 20001

NOTICE: The project that is the subject of this report was approved by the Governing Board of the National Research Council, whose members are drawn from the councils of the National Academy of Sciences, the National Academy of Engineering, and the Institute of Medicine. The members of the committee responsible for the report were chosen for their special competences and with regard for appropriate balance.

This study was supported by the National Oceanic and Atmospheric Administration under contract number WC133R-11-CQ-0048, TO#3, the National Science Foundation under grant number EAR-1305802, the United States intelligence community, and the National Academies. Any opinions, findings, and conclusions, or recommendations expressed in this material are those of the author(s) and do not necessarily reflect the views of the sponsoring agencies or any of their subagencies.

International Standard Book Number-13: 978-0-309-28773-9
International Standard Book Number-10: 0-309-28773-1
Library of Congress Control Number: 2013957745

Additional copies of this report are available for sale from the National Academies Press, 500 Fifth Street, NW, Keck 360, Washington, DC 20001; (800) 624-6242 or (202) 334-3313; http://www.nap.edu/.

Printed in the United States of America

THE NATIONAL ACADEMIES

Advisers to the Nation on Science, Engineering, and Medicine

The **National Academy of Sciences** is a private, nonprofit, self-perpetuating society of distinguished scholars engaged in scientific and engineering research, dedicated to the furtherance of science and technology and to their use for the general welfare. Upon the authority of the charter granted to it by the Congress in 1863, the Academy has a mandate that requires it to advise the federal government on scientific and technical matters. Dr. Ralph J. Cicerone is president of the National Academy of Sciences.

The **National Academy of Engineering** was established in 1964, under the charter of the National Academy of Sciences, as a parallel organization of outstanding engineers. It is autonomous in its administration and in the selection of its members, sharing with the National Academy of Sciences the responsibility for advising the federal government. The National Academy of Engineering also sponsors engineering programs aimed at meeting national needs, encourages education and research, and recognizes the superior achievements of engineers. Dr. C. D. Mote, Jr., is president of the National Academy of Engineering.

The **Institute of Medicine** was established in 1970 by the National Academy of Sciences to secure the services of eminent members of appropriate professions in the examination of policy matters pertaining to the health of the public. The Institute acts under the responsibility given to the National Academy of Sciences by its congressional charter to be an adviser to the federal government and, upon its own initiative, to identify issues of medical care, research, and education. Dr. Harvey V. Fineberg is president of the Institute of Medicine.

The **National Research Council** was organized by the National Academy of Sciences in 1916 to associate the broad community of science and technology with the Academy's purposes of furthering knowledge and advising the federal government. Functioning in accordance with general policies determined by the Academy, the Council has become the principal operating agency of both the National Academy of Sciences and the National Academy of Engineering in providing services to the government, the public, and the scientific and engineering communities. The Council is administered jointly by both Academies and the Institute of Medicine. Dr. Ralph J. Cicerone and Dr. C. D. Mote, Jr., are chair and vice chair, respectively, of the National Research Council.

www.national-academies.org

COMMITTEE ON UNDERSTANDING AND MONITORING ABRUPT CLIMATE CHANGE AND ITS IMPACTS

JAMES W.C. WHITE (*Chair*), University of Colorado, Boulder
RICHARD B. ALLEY, Pennsylvania State University, University Park
DAVID E. ARCHER, University of Chicago, IL
ANTHONY D. BARNOSKY, University of California, Berkeley
JONATHAN FOLEY, University of Minnesota, Saint Paul
RONG FU, University of Texas, Austin
MARIKA M. HOLLAND, National Center for Atmospheric Research, Boulder, CO
M. SUSAN LOZIER, Duke University, Durham, NC
JOHANNA SCHMITT, University of California, Davis
LAURENCE C. SMITH, University of California, Los Angeles
GEORGE SUGIHARA, University of California, San Diego
DAVID W. J. THOMPSON, Colorado State University, Fort Collins
ANDREW J. WEAVER, University of Victoria, British Columbia
STEVEN C. WOFSY, Harvard University, Cambridge, MA

NRC Staff

EDWARD DUNLEA, Senior Program Officer
CLAUDIA MENGELT, Senior Program Officer
AMANDA PURCELL, Research Associate
ROB GREENWAY, Program Associate

Preface

Climate is changing, forced out of the range of the last million years by levels of carbon dioxide and other greenhouse gases not seen in Earth's atmosphere for a very long time. Lacking action by the world's nations, it is clear that the planet will be warmer, sea level will rise, and patterns of rainfall will change. But the future is also partly uncertain—there is considerable uncertainty about *how* we will arrive at that different climate. Will the changes be gradual, allowing natural systems and societal infrastructure to adjust in a timely fashion? Or will some of the changes be more abrupt, crossing some threshold or "tipping point" to change so fast that the time between when a problem is recognized and when action is required shrinks to the point where orderly adaptation is not possible?

A study of Earth's climate history suggests the inevitability of "tipping points"—thresholds beyond which major and rapid changes occur when crossed—that lead to abrupt changes in the climate system. The history of climate on the planet—as read in archives such as tree rings, ocean sediments, and ice cores—is punctuated with large changes that occurred rapidly, over the course of decades to as little as a few years. There are many potential tipping points in nature, as described in this report, and many more that we humans create in our own systems. The current rate of carbon emissions is changing the climate system at an accelerating pace, making the chances of crossing tipping points all the more likely. The seminal 2002 National Academy Report, *Abrupt Climate Changes: Inevitable Surprises* (still required reading for anyone with a serious interest in our future climate) was aptly named: surprises are indeed inevitable. The question is now whether the surprises can be anticipated, and the element of surprise reduced. That issue is addressed in this report.

Scientific research has already helped us reduce this uncertainty in two important cases; potential abrupt changes in ocean deep water formation and the release of carbon from frozen soils and ices in the polar regions were once of serious near-term concern are now understood to be less imminent, although still worrisome as slow changes over longer time horizons. In contrast, the potential for abrupt changes in ecosystems, weather and climate extremes, and groundwater supplies critical for agriculture now seem more likely, severe, and imminent. And the recognition that a gradually changing climate can push both natural systems, as well as human systems, across tipping points has grown over the past decade. This report addresses both abrupt climate changes in the physical climate system, and abrupt climate impacts that occur in human and natural systems from a steadily changing climate.

In addition to a changing climate, multiple other stressors are pushing natural and human systems toward their limits, and thus become more sensitive to small perturbations that can trigger large responses. Groundwater aquifers, for example, are being depleted in many parts of the world, including the southeast of the United States. Groundwater is critical for farmers to ride out droughts, and if that safety net reaches an abrupt end, the impact of droughts on the food supply will be even larger.

Must abrupt changes always be surprises? Certainly not. As knowledge of the tipping points in natural and human systems improves, an early warning system can be developed. Careful and vigilant monitoring, combined with a constantly improving scientific understanding of the climate system, can help society to anticipate major changes before they occur. But it is also important to carefully and vigilantly catalog the assets at risk—societies cannot protect everything and will need to prioritize, and without an understanding of what could be lost, such as coastal infrastructure to rising seas, for example, intelligent decisions about what to protect first cannot be made.

Can all tipping points be foreseen? Probably not. Some will have no precursors, or may be triggered by naturally occurring variability in the climate system. Some will be difficult to detect, clearly visible only after they have been crossed and an abrupt change becomes inevitable. Imagine an early European explorer in North America, paddling a canoe on the swift river. This river happens to be named Niagara, but the paddler does not know that. As the paddler approaches the Falls, the roar of the water goes from faint to alarming, and the paddler desperately tries to make for shore. But the water is too swift, the tipping point has already been crossed, and the canoe—with the paddler—goes over the Falls. This tipping point is certainly hard to anticipate, but is it inevitable? No. The tipping point in this case could have been detected by an early warning system (listening for the roar of a waterfall), but importantly, prudence was required. Sticking closer to shore, in other words taking some prudent precautions, could have saved the paddler. Precaution will help us today as well, as we face a changing climate, if we are prudent enough to exercise it. Key to this is the need to be watching and listening for the early warning signals.

I would like to commend the committee for their hard work, stimulating conversations, scientific expertise, and most importantly, willingness to think outside of the box and take a fresh look at this issue. To Richard Alley (he of the 2002 Report), David Archer, Tony Barnosky, Jon Foley, Rong Fu, Marika Holland, Susan Lozier, Annie Schmitt, Larry Smith, George Sugihara, David Thompson, The Honorable Andrew Weaver and Steve Wofsy, I owe great thanks and heaps of praise. This report was an adventure, and I could not have asked for better travelling companions. The staff of the National Research Council, those heroes behind the scenes, worked tirelessly to make this study a success. Rob Greenway and Amanda Purcell spent countless hours pulling together text and organizing meetings. Claudia Mengelt contributed valuable discussions and tight editing. And Edward Dunlea simply did it all, keeping me and the rest of the crew on track, ensuring that we could indeed get this done in the time allotted (never enough, it seems) and making sure it all came together.

To the committee and staff, my deep, heartfelt thanks; intelligent, hard-working and industrious folks all. In fact, we have a planet full of intelligent, hard-working and industrious folks. Humans are capable of solving whatever problems nature throws at us, or that we create. But first we have to arm ourselves with information and then commit to using that information intelligently and wisely, and that, in a nutshell, is the message of this report.

Jim White, *Chair*
Committee on Understanding and Monitoring
Abrupt Climate Change and its Impacts

Acknowledgments

This report has been reviewed in draft form by individuals chosen for their diverse perspectives and technical expertise, in accordance with procedures approved by the National Research Council's (NRC's) Report Review Committee. The purpose of this independent review is to provide candid and critical comments that will assist the institution in making its published report as sound as possible and to ensure that the report meets institutional standards for objectivity, evidence, and responsiveness to the study charge. The review comments and draft manuscript remain confidential to protect the integrity of the deliberative process. The committee wishes to thank the following individuals for their review of this report:

PETER BREWER, Monterey Bay Aquarium Research Institute, Moss Landing, CA
ANTHONY JANETOS, Boston University, Massachusetts
JAMES KIRTLEY, Massachusetts Institute of Technology, Cambridge, MA
TIMOTHY LENTON, University of Exeter, UK
MARK PAGANI, Yale University, New Haven, CT
J.R. ANTHONY PEARSON, Schlumberger Cambridge Research, Cambridge, UK
DOROTHY PETEET, NASA Goddard / Columbia LDEO, Palisades, NY
ROGER PULWARTY, National Oceanic and Atmospheric Administration, Boulder, CO
PETER RHINES, University of Washington, Seattle
RICHARD L. SMITH, University of North Carolina, Chapel Hill
WILLIAM TRAVIS, University of Colorado, Boulder
GEORGE WOODWELL, Woods Hole Research Center, Massachusetts

Although the reviewers listed above have provided constructive comments and suggestions, they were not asked to endorse the views of the committee, nor did they see the final draft of the report before its release. The review of this report was overseen by **Marc Levy**, Columbia University, Palisades, NY, and **Warren M. Washington**, National Center for Atmospheric Research, Boulder, CO, appointed by the NRC Report Review Committee, who were responsible for making certain that an independent examination of this report was carried out in accordance with institutional procedures and that all review comments were carefully considered. Responsibility for the final content of this report rests entirely with the authoring panel and the institution.

In addition, the committee would like to thank the following individuals for their helpful discussions throughout the study process: Waleed Abdalati, Peter Brewer, Wally Broecker, Edward Dlugokencky, Ian Eismann, James Famiglietti, Jennifer Francis, Sherri Goodman, James Hansen, Paul Harnik, Stephen Jackson, Ian Joughin, David Lobell, Keith Moore, Ray Pierrehumbert, Lorenzo Polvani, William Reeburgh, Gabriel Vecchi, and Scott Wing.

Contents

Summary

Levels of carbon dioxide and other greenhouse gases in Earth's atmosphere are exceeding levels recorded in the past millions of years, and thus climate is being forced beyond the range of the recent geological era. Lacking concerted action by the world's nations, it is clear that the future climate will be warmer, sea levels will rise, global rainfall patterns will change, and ecosystems will be altered.

However, there is still uncertainty about *how* we will arrive at that future climate state. Although many projections of future climatic conditions have predicted steadily changing conditions giving the impression that communities have time to gradually adapt, for example, by adopting new agricultural practices to maintain productivity in hotter and drier conditions, or by organizing the relocation of coastal communities as sea level rises, the scientific community has been paying increasing attention to the possibility that at least some changes will be abrupt, perhaps crossing a threshold or "tipping point" to change so quickly that there will be little time to react. This concern is reasonable because such abrupt changes—which can occur over periods as short as decades, or even years—have been a natural part of the climate system throughout Earth's history. The paleoclimate record—information on past climate gathered from sources such as fossils, sediment cores, and ice cores—contains ample evidence of abrupt changes in Earth's ancient past, including sudden changes in ocean and air circulation, or abrupt extreme extinction events. One such abrupt change was at the end of the Younger Dryas, a period of cold climatic conditions and drought in the north that occurred about 12,000 years ago. Following a millennium-long cold period, the Younger Dryas abruptly terminated in a few decades or less and is associated with the extinction of 72 percent of the large-bodied mammals in North America.

Some abrupt climate changes are already underway, including the rapid decline of Arctic sea ice over the past decade due to warmer polar temperatures. In addition there are many parts of the climate system that have been thought to be possibly prone to near-future abrupt change that would trigger significant impacts at the regional and global scale. For some of these potential changes, current scientific understanding is insufficient to say with certainty how significant the threat is. In other cases, scientific research has advanced sufficiently that it is possible to assess the likelihood, for example the probability of a rapid shutdown of the Atlantic Meridional Overturning Circulation (AMOC) within this century is now understood to be low.

In addition to abrupt changes within the climate system itself, gradual climate changes can cross thresholds in both natural systems and human systems. For example, as air and water temperatures rise, some species, such as the mountain pika or some ocean corals, will no longer be able to survive in their current habitats and will be forced to relocate or rapidly adapt. Those populations that cannot do so quickly enough will be in danger of extinction. In addition, human infrastructure is built with certain expectations of useful life expectancy, but even gradual climate changes may trigger abrupt thresholds in their utility, such as rising sea levels surpassing sea walls or thawing permafrost destabilizing pipelines, buildings, and roads.

Climate is not the only stressor on the Earth system—other factors, including resource depletion and ever-growing human consumption and population, are exerting enormous pressure on nature's and society's resilience to sudden changes. Understanding the potential risks posed by both abrupt climate changes and the abrupt impacts resulting from gradual climate change is a crucial piece in advancing the ability of society to cope with changes in the Earth system. Better scientific understanding and improved ability to simulate the abrupt impacts of climate change would help researchers and policymakers with a comprehensive risk assessment. This report, sponsored by the US intelligence community, the National Oceanic and Atmospheric Administration, the National Science Foundation, and the National Academies, examines current knowledge about the likelihood and timing of potential abrupt changes, discusses the need for developing an abrupt change early warning system to help anticipate major changes before they occur, and identifies the gaps in the scientific understanding and monitoring capabilities (the full Statement of task can be found in Chapter 1).

STATE OF KNOWLEDGE ON ABRUPT IMPACTS OF CLIMATE CHANGE

This study differs from previous treatments of abrupt changes by discussing both the abrupt changes in the physical climate system (hereafter called "abrupt climate change"), as well as the abrupt changes in the physical, biological, or human systems that result from steadily changing aspects of the climate system (hereafter referred to as "abrupt climate impacts"). This report focuses on abrupt climate changes and abrupt climate impacts that have (or were thought to possibly have) the potential to severely affect the physical climate system, natural systems, or human systems, often affecting multiple interconnected areas of concern. The primary timescale of concern is years to decades. A key characteristic of these changes is that they can come faster than expected, planned, or budgeted for, forcing more reactive, rather than proactive, modes of behavior.

Careful and vigilant monitoring, combined with a constantly improving scientific understanding of the climate system, would help society anticipate major changes before they occur. With this goal in mind, the report's authoring committee summarized the state of knowledge about potential abrupt changes in Table S.1. This table includes potential abrupt changes to the ocean, atmosphere, ecosystems, and high-latitude regions that are judged to meet the above criteria. For each abrupt change, the Committee examined the available evidence of potential impact and likelihood. Some abrupt changes are likely to occur within this century—making these changes of most concern for near-term societal decision making and a priority for research. In other cases, there are still large scientific uncertainties about the likelihood of a potential abrupt change, highlighting the need for further research in these areas. Finally, recent data has revealed that some abrupt changes, widely discussed in the scientific literature because they were once identified as possible threats, are no longer considered likely during this century. This illustrates how focused efforts to study critical climate change mechanisms can also assuage societal concern about potential abrupt changes, in addition to identifying them.

Abrupt Changes Already Underway

The abrupt changes that are already underway are of most immediate concern for societal decisions. These include the disappearance of late-summer Arctic sea ice and increases in extinction rates of marine and terrestrial species.

Disappearance of Late-Summer Arctic Sea Ice

Recent dramatic changes in the extent and thickness of the ice that covers the Arctic sea have been well documented. Satellite data for late summer (September) sea ice extent show natural variability around a clearly declining long-term trend (Figure S.1). This rapid reduction in Arctic sea ice already qualifies as an abrupt change with substantial decreases in ice extent occurring within the past several decades. Projections from climate models suggest that ice loss will continue in the future, with the full disappearance of late-summer Arctic sea ice possible in the coming decades.

The impacts of rapid decreases in Arctic sea ice are likely to be considerable. More open water conditions during summer would have potentially large and irreversible effects on various components of the Arctic ecosystem, including disruptions in the marine food web, shifts in the habitats of some marine mammals, and erosion of vulnerable coastlines. Because the Arctic region interacts with the large-scale circu-

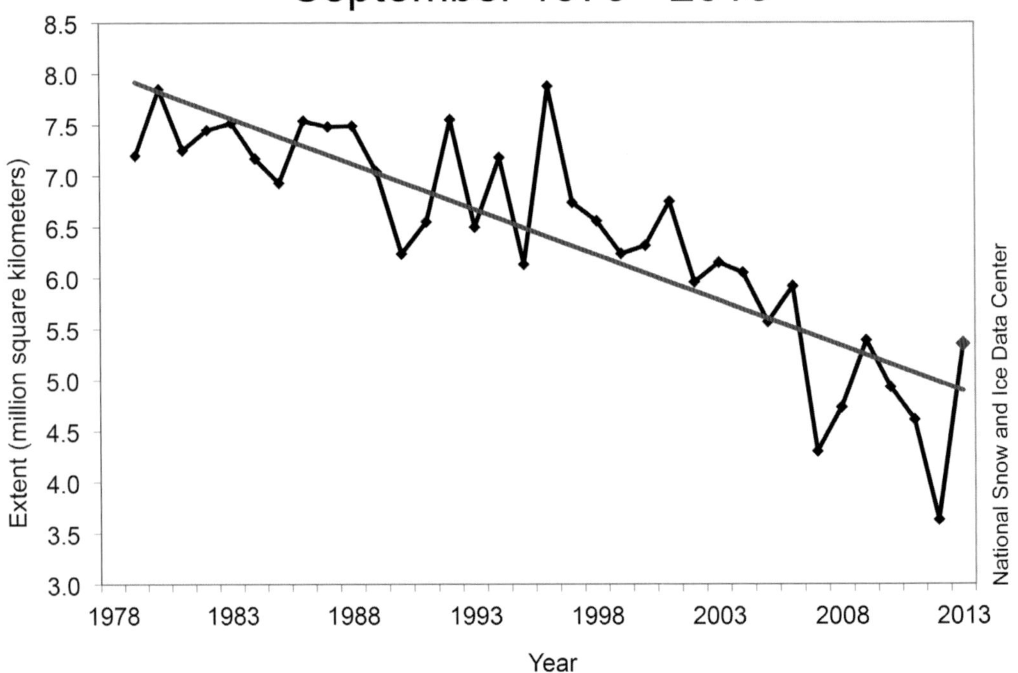

FIGURE S.1 The chart above shows the time series of Arctic sea ice extent each September from 1979 to 2013 as derived from satellite data. Late-summer Arctic sea ice extent has shown a substantial decrease since the satellite data record began in 1979, in particular the most recent seven summers have shown much lower sea ice cover. Source: National Snow and Ice Data Center, http://nsidc.org/arcticseaicenews/.

lation systems of the ocean and atmosphere, changes in the extent of sea ice could cause shifts in climate and weather around the northern hemisphere. The Arctic is also a region of increasing economic importance for a diverse range of stakeholders, and reductions in Arctic sea ice will bring new legal and political challenges as navigation routes for commercial shipping open and marine access to the region increases for offshore oil and gas development, tourism, fishing and other activities.

Understanding and predicting future changes in Arctic sea ice will require maintained and expanded observations of sea ice thickness and extent, including satellite-based measurements. Information on Arctic Ocean conditions may provide insight on the potential for rapid sea ice loss, yet only limited observations of Arctic Ocean conditions currently exist. In addition to observations, improved modeling of sea ice, including within global and regional climate models, is needed to better forecast Arctic sea ice changes and their impacts.

Increases in Extinction Threat for Marine and Terrestrial Species

The rate of climate change now underway is probably as fast as any warming event in the past 65 million years, and it is projected that its pace over the next 30 to 80 years will continue to be faster and more intense. These rapidly changing conditions make survival difficult for many species. Biologically important climatic attributes—such as number of frost-free days, length and timing of growing seasons, and the frequency and intensity of extreme events (such as number of extremely hot days or severe storms)—are changing so rapidly that some species can neither move nor adapt fast enough (Figure S.2).

Specific examples of species at risk for physiological reasons include mountain species such as pikas and endemic Hawaiian silverswords, which are restricted to cool temperatures at high altitudes. Species like polar bears are at risk because they depend on sea ice to facilitate their hunting of seals and Arctic sea ice conditions are changing rapidly. Other species are prone to extinction as changing climate causes their habitats to alter such that growth, development, or reproduction of constituent individuals are inhibited.

The distinct risks of climate change exacerbate other widely recognized and severe extinction pressures, especially habitat destruction, competition from invasive species, and unsustainable exploitation of species for economic gain, which have already elevated extinction rates to many times above background rates. If unchecked, habitat destruction, fragmentation, and over-exploitation, even without climate change, could result in a mass extinction within the next few centuries equivalent in magnitude to the one that wiped out the dinosaurs. With the ongoing pressures of climate change, comparable levels of extinction conceivably could occur before the year 2100; indeed, some models show a crash of coral reefs from climate change alone as early as 2060 under certain scenarios.

Loss of a species is permanent and irreversible, and has both economic impacts and ethical implications. The economic impacts derive from loss of ecosystem services, revenue, and jobs, for example in the fishing, forestry, and ecotourism industries. Ethical implications include the permanent loss of irreplaceable species and ecosystems as the current generation's legacy to the next generation.

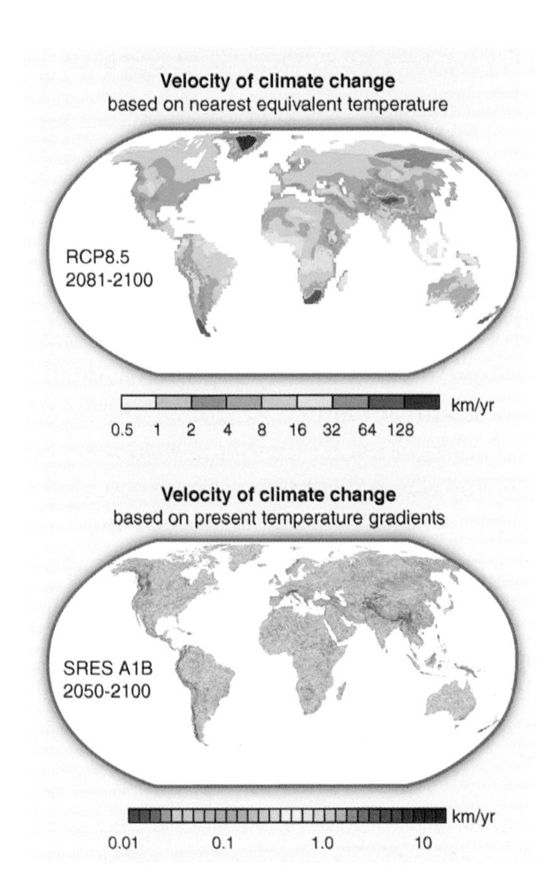

FIGURE S.2 As temperatures rise, populations of many species will have to move to new habitats to find suitable food, water, and shelter. The colors on these maps show how fast individuals in a species will have to move across the landscape in order to track the mean temperature that now characterizes the places where they live. The figure shows two methods of calculating the velocity of climate change for different time periods at the end of this century. The top panel shows the velocity in terms of nearest equivalent temperature, i.e., the climate change velocity in the CMIP5 RCP8.5 ensemble, calculated by identifying the closest location (to each grid point) with a future annual temperature that is similar to the baseline annual temperature. The lower panel expresses velocity as change in present temperature gradients calculated by using the present temperature gradient at each location and the trend in temperature projected by the CMIP3 ensemble in the SRES A1B scenario. Source: Diffenbaugh and Field, 2013.

Research on species extinctions is in many ways still at a nascent stage of discovery. Prominent research questions at this time include identifying which species in which ecosystems are most at risk, identifying which species extinctions would precipitate inordinately large ecological cascades that would lead to further extinctions, and assessing the impact of climate-induced changes in seasonal timing and species interactions on extinction rates.

Abrupt Changes of Unknown Probability

Destabilization of the West Antarctic Ice Sheet

The volume of ice sheets is controlled by the net balance between mass gained (from snowfall that turns to ice) and mass lost (from iceberg calving and the runoff of meltwater from the ice sheet). Scientists know with high confidence from paleo-climate records that during the planet's cooling phase, water from the ocean is traded for ice on land, lowering sea level by tens of meters or more, and during warming phases, land ice is traded for ocean water, raising sea level, again by tens of meters and more. The rates of ice and water loss from ice stored on land directly affect the speed of sea level rise, which in turn directly affects coastal communities. Of greatest concern among the stocks of land ice are those glaciers whose bases are well below sea level, which includes most of West Antarctica, as well as smaller parts of East Antarctica and Greenland. These glaciers are sensitive to warming oceans, which help to thermally erode their base, as well as rising sea level, which helps to float the ice, further destabilizing them. Accelerated sea level rise from the destabilization of these glaciers, with sea level rise rates several times faster than those observed today, is a scenario that has the potential for very serious consequences for coastal populations, but the probability is currently not well known, but probably low.

Research to understand ice sheet dynamics is particularly focused on the boundary between the floating ice and the grounded ice, usually called the grounding line (see Figure S.3). The exposed surfaces of ice sheets are generally warmest on ice shelves, because these sections of ice are at the lowest elevation, furthest from the cold central region of the ice mass and closest to the relatively warmer ocean water. Locations where meltwater forms on the ice shelf surface can wedge open crevasses and cause ice-shelf disintegration—in some cases, very rapidly.

Because air carries much less heat than an equivalent volume of water, physical understanding indicates that the most rapid melting of ice leading to abrupt sea-level rise is restricted to ice sheets flowing rapidly into deeper water capable of melting ice rapidly and carrying away large volumes of icebergs. In Greenland, such deep water contact with ice is restricted to narrow bedrock troughs where friction between ice and fjord walls limits discharge. Thus, the Greenland ice sheet is not expected to destabilize rapidly within this century. However, a large part of the West Antarctic

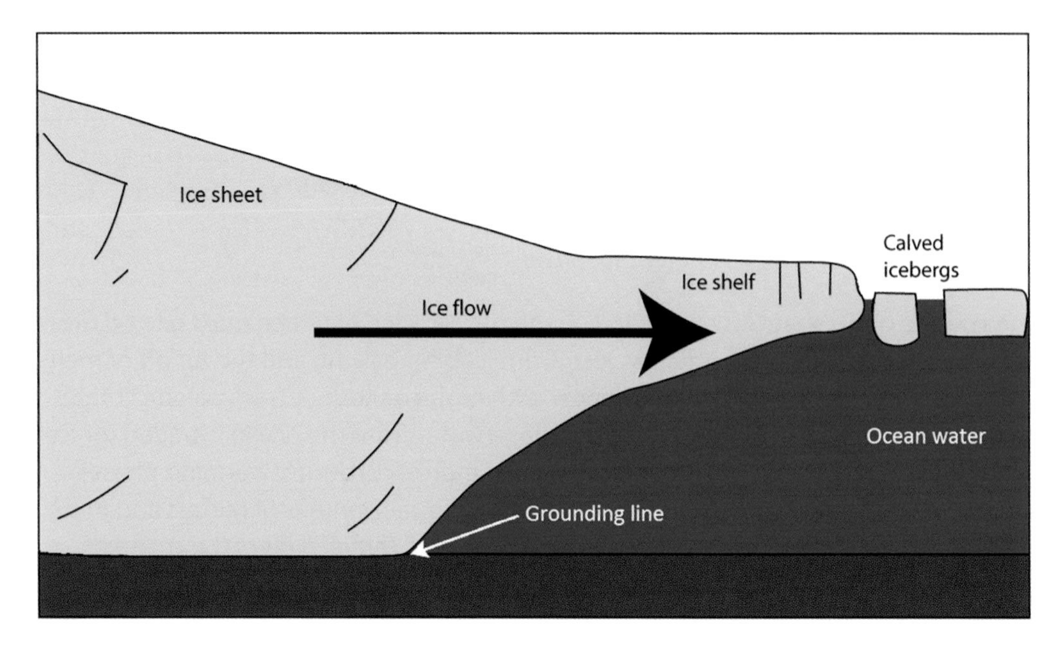

FIGURE S.3 The grounding line is the boundary between floating ice and grounded ice. SOURCE: Adapted from www.AntarcticGlaciers.org by Bethan Davies. Used with permission.

Ice Sheet (WAIS), representing 3–4 m of potential sea-level rise, is capable of flowing rapidly into deep ocean basins. Because the full suite of physical processes occurring where ice meets ocean is not included in comprehensive ice-sheet models, it remains possible that future rates of sea-level rise from the WAIS are underestimated, perhaps substantially. Improved understanding of key physical processes and inclusion of them in models, together with improved projections of changes in the surrounding ocean, are required to notably reduce uncertainties and to better quantify worst-case scenarios. Because large uncertainties remain, the Committee judges an abrupt change in the WAIS within this century to be plausible, with an unknown although probably low probability.

Abrupt Changes Unlikely to Occur This Century

Some abrupt changes that have been widely discussed in the literature because they were previously considered to be potential threats with poorly known probability. More recent research findings have shown that they may be less likely to occur within this century than previously considered possible. These include disruption to the

Atlantic Meridional Overturning Circulation (AMOC) and potential abrupt changes of high-latitude methane sources (permafrost soil carbon and ocean methane hydrates). Although the Committee judges the likelihood of an abrupt change within this century to be low for these processes, should they occur even next century or beyond, there would likely be severe impacts. Furthermore, gradual changes associated with these processes can still lead to consequential changes. Thus, they merit further study.

Disruption to the Atlantic Meridional Overturning Circulation (AMOC)

The AMOC is the ocean circulation pattern that involves the northward flow of warm near-surface waters into the northern North Atlantic and Nordic Seas, and the southward flow at depth of the cold dense waters formed in those high latitude regions. This circulation pattern plays a critical role in the global transport of oceanic heat, salt, and carbon. Paleoclimate evidence of temperature and other changes recorded in North Atlantic Ocean sediments, Greenland ice cores and other archives suggest that the AMOC abruptly shut down and restarted in the past—possibly triggered by large pulses of glacial meltwater or gradual meltwater supplies crossing a threshold—raising questions about the potential for abrupt change in the future.

Despite these concerns, recent climate and Earth system model simulations indicate that the AMOC is currently stable in the face of likely perturbations, and that an abrupt change will not occur in this century. This is a robust result across many different models, and one that eases some of the concerns about future climate change.

However, it is important keep a close watch on this system, to make observations of the North Atlantic to monitor how the AMOC responds to a changing climate, for reasons including the likelihood that slow changes will have real impacts, and to update the understanding of the slight possibility of a major event. One example of a monitoring effort began in 2004 when the U.K./U.S. RAPID/MOCHA array (Rapid Climate Change—Meridional Overturning Circulation and Heatflux Array) was deployed at 26.5°N to provide the first continuous measurement of the AMOC in the North Atlantic. Data from this array, which consists of instruments deployed from the North American continent to the west coast of Africa, has revealed key features of the AMOC and its variability. However, to understand the linkage between high-latitude climate change and AMOC variability, investigate the differences between South and North Atlantic AMOC variability, and to ground truth models of the AMOC system, measurements at other latitudes (currently being planned) are needed. Finally, to make a clear assessment of the AMOC's response to anthropogenic climate change, it is expected that a multi-decadal observing system will be necessary.

Potential Abrupt Changes due to High-Latitude Methane

Large amounts of carbon are stored at high latitudes in potentially labile reservoirs such as permafrost soils and methane-containing ices called methane hydrate or clathrate, especially offshore in ocean marginal sediments. Owing to their sheer size, these carbon stocks have the potential to massively affect Earth's climate should they somehow be released to the atmosphere. An abrupt release of methane is particularly worrisome because methane is many times more potent than carbon dioxide as a greenhouse gas over short time scales. Furthermore, methane is oxidized to carbon dioxide in the atmosphere, representing another carbon dioxide pathway from the biosphere to the atmosphere.

According to current scientific understanding, Arctic carbon stores are poised to play a significant amplifying role in the century-scale buildup of carbon dioxide and methane in the atmosphere, but are unlikely to do so abruptly, i.e., on a timescale of one or a few decades. Although comforting, this conclusion is based on immature science and sparse monitoring capabilities. Basic research is required to assess the long-term stability of currently frozen Arctic and sub-Arctic soil stocks, and of the possibility of increasing the release of methane gas bubbles from currently frozen marine and terrestrial sediments, as temperatures rise.

Summary of Abrupt Climate Changes and Abrupt Climate Impacts

In addition to the abrupt changes described in the sections above, the Committee examined a number of other possible changes. These included sea level rise due to thermal expansion or ice sheet melting (except WAIS—see above), decrease in ocean oxygen (expansion in oxygen minimum zones (OMZs)), changes to patterns of climate variability, changes in heat waves and extreme precipitation events (droughts/floods/hurricanes/major storms), disappearance of winter Arctic sea ice (distinct from late-summer Arctic sea ice—see above), and rapid state changes in ecosystems, species range shifts, and species boundary changes. Table S.1 summarizes the current knowledge of these various processes and identifies key future research and monitoring needs. This research promises to continue to help distinguish the more serious threats from the less likely ones.

ANTICIPATING SURPRISES

The abrupt climate changes and abrupt climate impacts discussed here present substantial risks to society and nature. The ability to anticipate what would otherwise be "surprises" in the climate system requires careful monitoring of climate conditions, improved models for projecting changes, and the interpretation and synthesis of scientific data using novel analysis techniques. In light of the importance of actionable information about the occurrence and impacts of abrupt changes, it is the Committee's judgment that action is urgently needed to improve society's ability to anticipate abrupt climate changes and impacts.

To address these needs the Committee recommends development of an Abrupt Change Early Warning System (ACEWS). Surprises in the climate system are inevitable: an early warning system could allow for the prediction and possible mitigation of such changes before their societal impacts are severe. Identifying key vulnerabilities can help guide efforts to increase resiliency and avoid large damages from abrupt change in the climate system, or in abrupt impacts of gradual changes in the climate system, and facilitate more informed decisions on the proper balance between mitigation and adaptation. With adequate scientific monitoring and study of these potential changes to the climate system, the probability that society can anticipate future abrupt climate changes and impacts will be substantially increased.

An ACEWS would be part of an overall risk management strategy, providing required information for hazard identification and risk assessment. In general, an ACEWS system would (1) identify and quantify social and natural vulnerabilities and ensure long-term, stable observations of key environmental and economic parameters through enhanced and targeted monitoring; (2) integrate new knowledge into numerical models for enhanced understanding and predictive capability; and (3) synthesize new learning and advance the understanding of the Earth system, taking advantage of collaborations and new analysis tools. The improved information could help identify vulnerabilities to assist in tailoring risk mitigation and preparedness efforts to ensure warnings result in the appropriate protective actions, with the ultimate goal to pre-empt catastrophes. Planning an ACEWS would benefit from leveraging the experience and knowledge gained as part of existing early warning programs such as the National Integrated Drought Information System (NIDIS) and Famine Early Warning System Network (FEWS NET). The Committee described several important aspects of a strategy to provide an effective Abrupt Change Early Warning System (ACEWS):

- ***Monitor key variables of abrupt change:*** Monitoring for an ACEWS should expand upon existing monitoring networks, protect and/or augment important networks that are currently in place, and develop new ones as needed (examples of specific monitoring needs are listed in Table S.1).
- ***Modeling to project future abrupt changes:*** A successful and adaptive ACEWS must consistently iterate between data collection, model testing and improve-

ment, and model predictions that suggest better data collection (examples of future modeling needs are listed in Table S.1).

- ***Synthesis of existing knowledge:*** A necessary part of an ACEWS is synthesizing knowledge to avoid the trap of data collection without continuing and evolving data analysis and model integration. This will require dedicated teams of researchers, improved collaborative networks, enhanced educational activities, and innovative tools for data analysis and modeling techniques.

To implement an ACEWS, it will be important to integrate the various components of the project, pay attention to stakeholder priorities, and build the ability to be flexible and adaptive. Thus, designing and implementing an ACEWS will need to be an iterative process that is revisited and refined as understanding of abrupt climate change, impacts, and social vulnerabilities evolves.

The organizational structure of an ACEWS could capitalize on existing programs, but there will be a need to capture the interconnectedness of climate and human systems. An ACEWS could eventually be run as a large, overarching program, but might better be started through coordination, integration, and expansion of existing and planned smaller programs. One possible mechanism to achieve this would be with a steering group that could provide efficient guidance. Such a steering committee could be made up of representatives of funding agencies, scientists, representatives of various user communities (including national security and interested businesses), and international partners, to name a subset of the possibilities. Subgroups or working groups may be able to bring focus to specific issues that require more attention as needed, e.g., water, food, or ecosystem services. A number of other interagency coordinating mechanisms exist that could assist in the planning and implementation of such a warning system. Whatever the mechanism, the committee does stress that coordination—to reduce duplication of efforts, maximize resources, and facilitate data and information sharing—is key to a successful ACEWS. The development of an ACEWS will need to be an ongoing process, one that goes beyond the scope of this report, and one that needs to include multiple stakeholders.

THE WAY FORWARD

Scientific understanding of abrupt changes in the physical climate system and abrupt impacts of climate change has steadily advanced over the past couple of decades. Owing to these scientific advances, some of the possible abrupt climate change mechanisms whose probability of occurrence was previously poorly known are now understood to be unlikely during this century—these include a sudden shutdown of the Atlantic Meridional Overturning Circulation, and a large and abrupt release of methane from thawing Arctic permafrost. However, concerns over the likelihood of other potential abrupt impacts of climate change—such as destabilization of the West Antarctic Ice Sheet and rapid increases in already-high rates of species extinctions—have intensified. It is important to note that such abrupt impacts can be suddenly triggered simply by continuing the present climate-change trajectory that humans are driving until "tipping points" are reached, as opposed to an abrupt change in the climate system itself.

The committee believes strongly that actions are needed to develop an ACEWS that serves to anticipate the possibilities of future abrupt changes and helps to reduce the potential consequences from such abrupt changes. Knowledge in this field is continuously advancing, and the implementation of such an early warning system will require additional collaborative research targeted at how to synthesize the various components in the most effective way. The proper design and implementation of an ACEWS will need to be an ongoing process and will require expertise from many different disciplines beyond just the physical sciences, as well as input from many different stakeholder groups. Providing a complete roadmap to a successful ACEWS was beyond the scope of this report, but the committee has outlined its initial thoughts on what would make such a system successful above. Much is known about the design, implementation, and sustainability of early warning systems that can be leveraged in addition what is described in this report.

Although there is still much to learn about abrupt climate change and abrupt climate impacts, to willfully ignore the threat of abrupt change could lead to more costs, loss of life, suffering, and environmental degradation. The time is here to be serious about the threat of tipping points so as to better anticipate and prepare ourselves for the inevitable surprises.

TABLE S.1 State of knowledge on potential candidate processes that might undergo abrupt change. These include both abrupt climate changes in the physical climate system and abrupt climate impacts of ongoing changes that, when certain thresholds are crossed, can cause abrupt impacts for society and ecosystems. The near term outlook for this century is highlighted as being of particular relevance for decision makers generally.

	Potential Abrupt Climate Change or Impact and Key Examples of Consequences	Current Trend	Near Term Outlook (for an Abrupt Change within This Century)	Long Term Outlook (for a Significant Change[1] after 2100)	Level of Scientific Understanding	Critical Needs (Research, Monitoring, etc.)
Abrupt Changes in the Ocean	**Disruption to Atlantic Meridional Overturning Circulation (AMOC)** • Up to 80 cm sea level rise in North Atlantic • Southward shift of tropical rain belts • Large disruptions to local marine ecosystems • Ocean and atmospheric temperature and circulation changes • Changes in ocean's ability to store heat and carbon	Trend not clearly detected	**Low**	High	Moderate	• Enhanced understanding of changes at high latitudes in the North Atlantic (e.g., warming and/or freshening of surface waters) • Monitoring of overturning at other latitudes • Enhanced understanding of drivers of AMOC variability
Abrupt Changes in the Ocean	**Sea level rise (SLR) from ocean thermal expansion** • Coastal inundation • Storm surges more likely to cause severe impacts	Moderate increase in sea level rise	**Low[2]**	High	High	• Maintenance and expansion of monitoring of sea level (tide gauges and satellite data), ocean temperature at depth, local coastal motions, and dynamic effects on sea level
Abrupt Changes in the Ocean	**Sea level rise from destabilization of WAIS ice sheets** • 3-4 m of potential sea level rise • Coastal inundation • Storm surges more likely to cause severe impacts	Losing ice to raise sea level	**Unknown but Probably Low**	Unknown	Low	• Extensive needs, including broad field, remote-sensing, and modeling research
Abrupt Changes in the Ocean	**Sea level rise from other ice sheets (including Greenland and all others, but not including WAIS loss)** • As much as 60m of potential sea level rise from all ice sheets • Coastal inundation • Storm surges more likely to cause severe impacts	Losing ice to raise sea level	**Low**	High	High for some aspects, Low for others	• Maintenance and expansion of satellite, airborne, and surface monitoring capacity, process studies, and modeling research

continued

TABLE S.1 Continued

	Potential Abrupt Climate Change or Impact and Key Examples of Consequences	Current Trend	Near Term Outlook (for an Abrupt Change within This Century)	Long Term Outlook (for a Significant Change[1] after 2100)	Level of Scientific Understanding	Critical Needs (Research, Monitoring, etc.)
...in the Ocean (cont.)	**Decrease in ocean oxygen (expansion in oxygen minimum zones [OMZs])** • Threats to aerobic marine life • Release of nitrous oxide gas—a potent greenhouse gas—to the atmosphere	Trend not clearly detected	**Moderate**	High	Low to Moderate	• Expanded and standardized monitoring of ocean oxygen content, pH, and temperature • Improved understanding and modeling of ocean mixing • Improved understanding of microbial processes in OMZs
Abrupt Changes in the Atmosphere	**Changes to patterns of climate variability (e.g., ENSO, annular modes)** • Substantial surface weather changes throughout much of extratropics if the extratropical jetstreams were to shift abruptly	Trends not detectable for most patterns of climate variability Exception is southern annular mode—detectable poleward shift of middle latitude jetstream	**Low**	Moderate	Low to Moderate	• Maintaining continuous records of atmospheric pressure and temperatures from both in-situ and remotely sensed sources • Assessing robustness of circulation shifts in individual ensemble members in climate change simulations • Developing theory on circulation response to anthropogenic forcing
	Increase in intensity, frequency, and duration of heat waves • Increased mortality • Decreased labor capacity • Threats to food and water security	Detectable increasing trends	**Moderate** (Regionally variable, dependent on soil moisture)	High	High	• Continued progress on understanding climate dynamics • Increased focus on risk assessment and resilience
	Increase in frequency and intensity of extreme precipitation events (droughts/floods/hurricanes/major storms) • Mortality risks • Infrastructure damage • Threats to food and water security • Potential for increased conflict	Increasing trends for floods Trends for drought and hurricanes not clear	**Moderate**	Moderate to High	Low to Moderate	• Continued progress on understanding climate dynamics • Increased focus on risk assessment and resilience

TABLE S.1 Continued

	Potential Abrupt Climate Change or Impact and Key Examples of Consequences	Current Trend	Near Term Outlook (for an Abrupt Change within This Century)	Long Term Outlook (for a Significant Change[1] after 2100)	Level of Scientific Understanding	Critical Needs (Research, Monitoring, etc.)
Abrupt Changes at High Latitudes	**Increasing release of carbon stored in soils and permafrost** • Amplification of human-induced climate change[3]	Neutral trend to small trend in increasing soil carbon release	**Low**	High	Moderate[4]	• Improved models of hydrology/cryosphere interaction and ecosystem response • Greater study of role of fires in rapid carbon release • Expanded borehole temperature monitoring networks • Enhanced satellite and ground-based monitoring of atmospheric methane concentrations at high latitudes
Abrupt Changes at High Latitudes	**Increasing release of methane from ocean methane hydrates** • Amplification of human-induced climate change	Trend not clearly detected	**Low[5]**	Moderate	Moderate[6]	• Field and model based characterization of the sediment column • Enhanced satellite and ground-based monitoring of atmospheric methane concentrations at high latitudes
Abrupt Changes at High Latitudes	**Late-summer Arctic sea ice disappearance** • Large and irreversible effects on various components of the Arctic ecosystem • Impacts on human society and economic development in coastal polar regions • Implications for Arctic shipping and resource extraction • Potential to alter large-scale atmospheric circulation and its variability	Strong trend in decreasing sea ice cover	**High**	Very high	High	• Enhanced Arctic observations, including atmosphere, sea ice, and ocean characteristics • Better monitoring and census studies of marine ecosystems • Improved large-scale models that incorporate the evolving state of knowledge
Abrupt Changes at High Latitudes	**Winter Arctic sea ice disappearance** • Same as late summer Arctic sea ice disappearance above, but more pronounced due to year-round lack of sea ice	Small trend (Decreasing but not disappearing)	**Low**	Moderate	High	• Same as late summer Arctic sea ice disappearance above

continued

TABLE S.1 Continued

	Potential Abrupt Climate Change or Impact and Key Examples of Consequences	Current Trend	Near Term Outlook (for an Abrupt Change within This Century)	Long Term Outlook (for a Significant Change[1] after 2100)	Level of Scientific Understanding	Critical Needs (Research, Monitoring, etc.)
Abrupt Changes in Ecosystems	**Rapid state changes in ecosystems, species range shifts, and species boundary changes** • Extensive habitat loss • Loss of ecosystem services • Threats to food and water supplies	Species range shifts significant; others not clearly detected	**Moderate**	High	Moderate	• Long term remote sensing and in-situ studies of key systems • Improved hydrological and ecological models
	Increases in extinctions of marine and terrestrial species • Loss of high percentage of coral reef ecosystems (already underway) • Significant percentage of land mammal, bird, and amphibian species extinct or endangered[7]	Species and population losses accelerating (Portion attributable to climate is uncertain)	**High**	Very high	Moderate	• Better understanding of how species interactions and ecological cascades might magnify extinctions intensity • Better understanding of how interactions between climate-caused extinctions and other extinction drivers (habitat fragmentation, overexploitation, etc.) multiply extinction intensity • Improved monitoring of key species

[1] Change could be either abrupt or non-abrupt.
[2] To clarify, the Committee assesses the near-term outlook that sea level will rise abruptly before the end of this century as Low; this is not in contradiction to the assessment that sea level will continue to rise steadily with estimates of between 0.26 and 0.82 m by the end of this century (IPCC, 2013).
[3] Methane is a powerful but short-lived greenhouse gas.
[4] Limited by ability to predict methane production from thawing organic carbon
[5] No mechanism proposed would lead to abrupt release of substantial amounts of methane from ocean methane hydrates this century.
[6] Limited by uncertainty in hydrate abundance in near-surface sediments, and fate of CH_4 once released
[7] Species distribution models (Thuiller et al., 2006) indicate between 10–40% of mammals now found in African protected areas will be extinct or critically endangered by 2080 as a result of modeled climate change. Analyses by Foden et al.(2013) and Ricke et al. (2013) suggest 41% of bird species, 66% of amphibian species, and between 61% and 100% of corals that are not now considered threatened with extinction will become threatened due to climate change sometime between now and 2100.

1

CHAPTER ONE

Introduction

The idea that Earth's climate could abruptly change in a drastic manner has been around for several decades. Early studies of ice cores showed that very large changes in climate could happen in a matter of a few decades or even years, for example, local to regional temperature changes of a dozen degrees or more, doubling or halving of precipitation rates, and dust concentrations changing by orders of magnitude (Dansgaard et al., 1989; Alley et al., 1993). In the last few decades, scientific research has advanced our understanding of abrupt climate change significantly. Some original fears have been allayed or now seem less ominous, but new ones have sprung up. Fresh reminders occur regularly that thresholds and tipping points exist not only in the climate system, but in other parts of the Earth system (Box 1.1).

What has become clearer recently is that the issue of abrupt change cannot be confined to a geophysical discussion of the climate system alone. The key concerns are not limited to large and abrupt shifts in temperature or rainfall, for example, but also extend to other systems that can exhibit abrupt or threshold-like behavior even in response to a gradually changing climate. The fundamental concerns with abrupt change include those of speed—faster changes leave less time for adaptation, either economically or ecologically—and of magnitude—larger changes require more adaptation and generally have greater impact. This report offers an updated look at the issue of abrupt climate change and its potential impacts, and takes the added step of considering not only abrupt changes to the climate system itself, but also abrupt impacts and tipping points that can be triggered by gradual changes in climate.

This examination of the impacts of abrupt change brings the discussion into the human realm, raising questions such as: Are there potential thresholds in society's ability to grow sufficient food? Or to obtain sufficient clean water? Are there thresholds in the risk to coastal infrastructure as sea levels rise? The spectrum of possibilities here is very wide, too wide to be fully covered in any single report. In practice, little is known about these and other possible abrupt changes. As such, this report lays out what is currently known about the risks, raises flags to point out potential threats, and proposes improved monitoring and warning schemes to help prepare us for both known and unknown abrupt changes. This report can be viewed as the current frame in an ongoing movie in which we grasp the basic plot, but we are not sure what plot twists lie ahead or even how the various characters are related. As scientific research and monitoring progresses, i.e., as we watch the movie and learn more about the key characters and

BOX 1.1 EXAMPLES OF RECENT ABRUPT CHANGE IN THE EARTH SYSTEM

Stratospheric Ozone Depletion

During the early 1970s, concerns arose in the scientific community that inputs of nitrogen oxides (known as "NOx") from a proposed fleet of supersonic aircraft flying in the stratosphere and of industrially produced halocarbon gases containing chlorine and bromine (CFCs or chlorofluorocarbons and chlorofluorobromocarbons) had the potential to deplete the amount of ozone in the stratosphere. Halogen oxide radicals were predicted to form from the degradation of halocarbons in the stratosphere. Intensive study of the stratosphere, extending more than a decade, confirmed the rising concentrations of CFCs and halons in the atmosphere, and of halogen oxide radicals in the stratosphere. International negotiations led to the signing of the Montreal Protocol in 1987, requiring a 50 percent reduction in CFCs and a 100 percent reduction in halon production by 2000 by the developed countries.

However, two years prior to the treaty, scientists learned that the column amount of ozone over Antarctica in the austral spring had been declining since the late 1960s, and it had been reduced by almost a factor of two by the mid-1980s (Farman et al., 1985); See Figure A. The continuous record of column ozone abundances measured at Halley Bay, Antarctica, showed

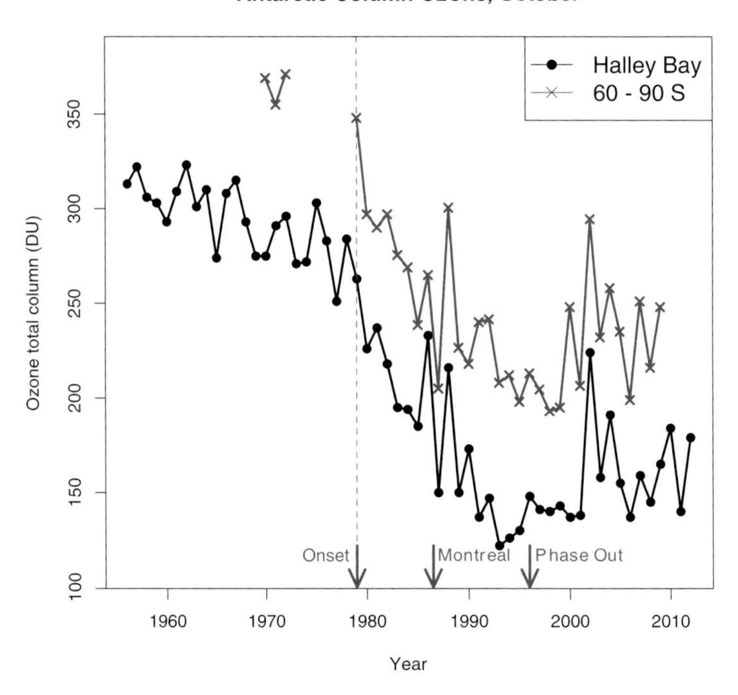

FIGURE A Total column ozone in Antarctica, at the Halley Bay station of the British Antarctic Survey (black) and averaged over the whole polar region of Antarctica (blue, from satellite data). (Adapted from WMO/UNEP [2010] plus data from the British Antarctic Survey [http://www.antarctica.ac.uk/met/jds/ozone/data/ZOZ5699.DAT, downloaded 26 April 2013].)

BOX 1.1 CONTINUED

that October ozone column amounts started to drift lower in the late 1960s and 1970s. Satellite records and measurements from other stations confirmed that this change was occurring on the continental scale of Antarctica. This was an abrupt change in the timescale of human activities, the scale of the whole polar region, but lack of continuity and rejection of data perceived to be anomalous prevented the detection of the change from space observations.

The Montreal Protocol was amended to require complete phase-out of most ozone-depleting CFCs by 1996 in developed countries and by 2010 in rest of the world. In addition, the Protocol was amended or adjusted multiple times to reduce emissions of all ozone-depleting substances. As a result of the Montreal Protocol and its amendments, stratospheric ozone is expected to return to its pre-1980 values as the atmospheric abundances of ozone-depleting substances decline in the coming decades. Column global ozone amounts prevalent in the early1970s are expected to be restored by the mid-21st century, although stratospheric cooling associated with changes in greenhouse gases will alter the trajectory of the restoration. The Antarctic ozone hole is expected to no longer occur towards the late 21st century, and this recovery is not expected to be influenced as much by climate change as the global ozone amounts (WMO/UNEP, 2010).

The Antarctic ozone hole represents an abrupt change to the Earth system. Although it is not specifically an abrupt climate change, for the purposes of this report, it is a recent example of the type an unforeseen global threshold event. The Antarctic ozone hole appeared within a few years after a threshold was crossed—when the concentrations of inorganic chlorine exceeded the concentration of nitrogen oxides in the lower altitudes of the polar stratosphere—and it affected a large portion of the globe. Thus, it exemplifies the scope and magnitude of the types of impacts that abrupt changes from human activities can have on the planet.

Bark Beetle Outbreaks

Bark beetles are a natural part of forested ecosystems, and infestations are a regular force of natural change. In the last two decades, though, the bark beetle infestations that have occurred across large areas of North America have been the largest and most severe in recorded history, killing millions of trees across millions of hectares of forest from Alaska to southern California (Bentz, 2008); see Figure B. Bark beetle outbreak dynamics are complex, and a variety of circumstances must coincide and thresholds must be surpassed for an outbreak to occur on a large scale. Climate change is thought to have played a significant role in these recent outbreaks by maintaining temperatures above a threshold that would normally lead to cold-induced mortality. In general, elevated temperatures in a warmer climate, particularly when there are consecutive warm years, can speed up reproductive cycles and increase the likelihood of outbreaks (Bentz et al., 2010). Similar to many of the issues described in this report, climate change is only one contributing factor to these types of abrupt climate impacts, with other human actions such as forest history and management also playing a role. There are also feedbacks to the climate system from these outbreaks, which represent an important mechanism by which climate change may undermine the ability of northern forests to take up and store atmospheric carbon (Kurz et al., 2008).

continued

BOX 1.1 CONTINUED

FIGURE B Photographs of a pine bark beetle and of a beetle-killed forest in the Colorado Rocky Mountains. Source: Top: Photo by Dion Manastyrski; Bottom: Photo from Anthony Barnosky.

how they interact, it is hoped that scientists and policymakers will learn to anticipate abrupt plot changes and surprises so that societies can be better prepared to handle them.

PREVIOUS DEFINITIONS OF ABRUPT CLIMATE CHANGE

As recently as the 1980s, the typical view of major climate change was one of slow shifts, paced by the changes in solar energy that accompany predictable variations in Earth's orbit around the sun over thousands to tens of thousands of years (Hays et al., 1976). While some early studies of rates of climate change, particularly during the last glacial period and the transition from glacial to interglacial climates, found large changes in apparently short periods of time (e.g., Coope et al., 1971), most of the paleoclimate records reaching back tens of thousands of years lacked the temporal resolution to resolve yearly to decadal changes. This situation began to change in the late 1980s as scientists began to examine events such as the climate transition that occurred at the end of the Younger Dryas about 12,000 years ago (e.g., Dansgaard et al., 1989) and the large swings in climate during the glacial period that have come to be termed "Dansgaard-Oescher events" ("D-O events;" named after two of the ice core scientists who first studied these phenomena using ice cores). At first these variations seemed to many to be too large and fast to be climatic changes, and it was only after they were found in several ice cores (e.g., Anklin et al., 1993; Grootes et al., 1993),[1] and in many properties (e.g., Alley et al., 1993), including greenhouse gases (e.g., Severinghaus and Brook, 1999) that they became widely accepted as real.

This perspective is important, as first definitions of abrupt climate change were tied directly to these D-O events, which themselves are defined by changes in temperature, precipitation rates, dust fallout, and concentrations of certain greenhouse gases. For this reason, previous reviews of abrupt change have tended to focus on the physical climate system, and the potential for abrupt changes and threshold behavior has been expressed primarily in climatic terms (key references listed in Box 1.2).

The first systematic review of abrupt climate change was by the National Research Council (*Abrupt Climate Change: Inevitable Surprises;* NRC, 2002). This study defined abrupt climate change as follows:

> "Technically, an abrupt climate change occurs when the climate system is forced to cross some threshold, triggering a transition to a new state at a rate determined by the

[1] http://www.gisp2.sr.unh.edu/.

BOX 1.2 PREVIOUS REPORTS ON ABRUPT CLIMATE CHANGE

Key references on the subject of abrupt climate change:

- *Abrupt Climate Change: Inevitable Surprises* (NRC, 2002)
- *IPCC Fourth Assessment Report: Climate Change 2007* (IPCC, 2007c)
- *Synthesis and Assessment Product 3.4: Abrupt Climate Change* (USCCSP, 2008)
- *Tipping Elements in the Earth's Climate System* (Lenton et al., 2008)
- *Climate and Social Stress: Implications for Security Analysis* (NRC, 2012a)
- *2013 National Climate Assessment* (National Climate Assessment and Development Advisory Committee, 2013)

> climate system itself and faster than the cause. Chaotic processes in the climate system may allow the cause of such an abrupt climate change to be undetectably small."

This early definition is critically important in two regards. First, it focuses on the climate system itself, a focus that remains widely used today. Second, it raises the possibility of thresholds or tipping points being forced or pushed by an undetectably small change in the cause of the shift. The 2002 report goes on to expand on its definition by placing abrupt climate change into a social context:

> "To use this definition in a policy setting or public discussion requires some additional context, ... because while many scientists measure time on geological scales, most people are concerned with changes and their potential impacts on societal and ecological time scales. From this point of view, an abrupt change is one that takes place so rapidly and unexpectedly that human or natural systems have difficulty adapting to it. Abrupt changes in climate are most likely to be significant, from a human perspective, if they persist over years or longer, are larger than typical climate variability, and affect sub-continental or larger regions. Change in any measure of climate or its variability can be abrupt, including change in the intensity, duration, or frequency of extreme events."

This expanded definition raised the issues of persistence, of changes being so large that they stand out above typical variability, and that changes in extremes, not just baselines, were considered to be abrupt climate changes. It also placed climate change into the context of *impacts* of those changes, and the change being considered abrupt if it exceeds the system's capacity to adapt.

In the subsequent years many papers were published on abrupt climate change, some with definitions more focused on time (e.g., Clark et al., 2002), and others on the relative speed of the causes and reactions. Overpeck and Cole (2006), for example, defined abrupt climate change as "a transition in the climate system whose duration is fast relative to the duration of the preceding or subsequent state." Lenton et al. (2008) formally introduced the concept of tipping point, defining abrupt climate change as:

> "We offer a formal definition, introducing the term "tipping element" to describe subsystems of the Earth system that are at least subcontinental in scale and can be switched—under certain circumstances—into a qualitatively different state by small perturbations. The tipping point is the corresponding critical point—in forcing and a feature of the system—at which the future state of the system is qualitatively altered."

In 2007, the Intergovernmental Panel on Climate Change Fourth Assessment report defined abrupt climate change as:

> "forced or unforced climatic change that involves crossing a threshold to a new climate regime (e.g., new mean state or character of variability), often where the transition time to the new regime is short relative to the duration of the regime."

In late 2008 a report of the U.S. Climate Change Science Program (USCCSP) was dedicated to the topic of abrupt climate change. The *Synthesis and Assessment Product 3.4: Abrupt Climate Change* (USCCSP, 2008) defined abrupt climate change as:

> "A large-scale change in the climate system that takes place over a few decades or less, persists (or is anticipated to persist) for at least a few decades, and causes substantial disruptions in human and natural systems."

This simple definition directly focuses attention on the impacts of change on natural and human systems and is important in that it directly combines the physical climate system with human impacts. As an increasingly interdisciplinary approach was taken to studying abrupt climate change, there was an accompanying evolution in thinking, expanding from abrupt changes in the physical climate system to include abrupt impacts from climate change.

More recently, *Climate and Social Stress: Implications for Security Analysis* (NRC, 2012b) examined the topic of climate change in the context of national security and briefly addressed the issue of abrupt climate change. They noted that events that did not meet the common criterion of a semi-permanent change in state could still force other systems into a permanent change, and thus qualify as an abrupt change. For example, a mega-drought may be followed by the return of normal precipitation rates, such that no baseline change occurred, but if that drought caused the collapse of a civilization, a permanent, abrupt change occurred in the system impacted by climate.

The 2002 NRC study introduced the important issue of gradual climate change causing abrupt responses in human or natural systems, noting "Abrupt impacts therefore have the potential to occur when gradual climatic changes push societies or ecosystems across thresholds and lead to profound and potentially irreversible impacts." The 2002 report also noted that "...the more rapid the forcing, the more likely it is that the resulting change will be abrupt on the time scale of human economies or global ecosystems" and "The major impacts of abrupt climate change are most likely to occur when economic or ecological systems cross important thresholds" (NRC, 2002). The 2012 NRC study embraced this issue more fully and expanded on the concept. The first part of their definition is straightforward:

> "Abrupt climate change is generally defined as occurring when some part of the climate system passes a threshold or tipping point resulting in a rapid change that produces a new state lasting decades or longer (Alley et al., 2003). In this case "rapid" refers to timelines of a few years to decades."

The second part of their definition echoes the 2002 report in emphasizing the role of abrupt responses to gradually changing forcing (emphasis added):

> "Abrupt climate change can occur on a regional, continental, hemispheric, or even global basis. *Even a gradual forcing of a system with naturally occurring and chaotic variability can cause some part of the system to cross a threshold, triggering an abrupt change.* Therefore, it is likely that gradual or monotonic forcings increase the probability of an abrupt change occurring."

DEFINITION OF ABRUPT CLIMATE CHANGE FOR THIS REPORT

The committee embraces the broader concept of abrupt climate change described in the 2002 NRC report and the definition from the 2012 *Climate and Social Stress* report, while expanding the scope of the definition further by considering *abrupt climate impacts*, as well as *abrupt climate changes* (Box 1.3). This distinction is critical, and represents a broadening of the focus from just the physical climate system itself to also encompass abrupt changes in the natural and human-built world that may be triggered by gradual changes in the physical climate system. Thus, the committee begins by defining that, for this report, the term "abrupt climate change" as being abrupt changes in the physical climate system, and the related term, "abrupt climate impacts," as being abrupt impacts resulting from climate change, even if the climate change itself is gradual (but reaches a threshold value that triggers an abrupt impact in a related system)

This definition of abrupt climate change also helps to set a time frame for what kinds of phenomena are considered in this report. Environmental changes occurring over timescales exceeding 100 years are not frequently considered in decision-making by the general public, private sector, or the government. For some, projected changes oc-

BOX 1.3 DEFINITION OF ABRUPT CLIMATE CHANGE FOR THIS REPORT

The subject of this report includes both the abrupt changes in the physical climate system (hereafter called "abrupt climate change") and abrupt impacts in the physical, biological, or human systems triggered by a gradually changing climate (hereafter called "abrupt climate impacts". These abrupt changes can affect natural or human systems, or both. The primary timescale of concern is years to decades. A key characteristic of these changes is that they can unfold faster than expected, planned for, or budgeted for, forcing a reactive, rather than proactive mode of behavior. These changes can propagate systemically, rapidly affecting multiple interconnected areas of concern.

curring over less than 100 years begin to raise questions related to inter-generational equity and can be viewed as a relevant time frame for certain policy settings. Changes occurring over a few decades, i.e., a generation or two, begin to capture the interest of most people because it is a time frame that is considered in many personal decisions and relates to personal memories. Also, at this time scale, changes and impacts can occur faster than the expected, stable lifetime of systems about which society cares. For example, the sizing of a new air conditioning system may not take into consideration the potential that climate change could make the system inadequate and unusable before the end of its useful lifetime (often 30 years or more). The same concept applies to other infrastructure, such as airport runways, subway systems, and rail lines. Thus, even if a change is occurring over several decades, and therefore might not at first glance seem "abrupt," if that change affects systems that are expected to function for an even longer period of time, the impact can indeed be abrupt when a threshold is crossed. "Abrupt" then, is relative to our "expectations," which for the most part come from a simple linear extrapolation of recent history, and "expectations" invoke notions of risk and uncertainty. In such cases, it is the cost associated with unfulfilled expectations that motivates discussion of abrupt change. Finally, changes occurring over one to a few years are abrupt, and for most people, would also be alarming if sufficiently large and impactful. In this report, the committee adopts the time frame for "abrupt" climate changes as years to decades.

The committee chose to focus their discussions of abrupt climate changes to those relevant to human society, including changes in the physical climate itself, and resulting changes to human expectations. Given our reliance on natural systems for ecosystem services, impacts to natural systems are of great concern to society as well. This consideration of unexpected impacts to societies and ecosystems broadens the discussion beyond the physics and chemistry of the climate system to include effects on humans and biota on local, regional, national, and international scales occurring over years to decades. This is a broad definition that could easily encompass too many topics to cover in one report, and in this report, the committee has attempted to steer clear of the temptation to craft a laundry list of topics. As the climate science community is in the early stages of examining many potential socioeconomic impacts, the discussion is thus necessarily limited in this report to those impacts for which there is good reason to suspect they are both abrupt and could actually occur.

There is a nascent but rapidly growing literature on the theory behind how abrupt transitions occur (Box 1.4). This research is also beginning to tackle the even harder question of how to anticipate abrupt transitions, across many disciplines and systems, a topic that this report returns to in more depth in Chapter 4.

BOX 1.4 MECHANISMS OF ABRUPT CHANGE

Shocks or sudden events in the environment have often been classified into categories based on duration: (1) large temporary disturbances (e.g., earthquakes, hurricanes, tsunamis); and (2) shifts in long-term behavior (e.g., El Niño events, glacial cycles) (Lenton, 2013). However, both of these categories are really different aspects of the same fundamental phenomenon, a change in the system dynamics from the "normal" behavior.

Although much is unknown about the mechanisms that can result in abrupt changes, some examples where there has been progress include positive feedbacks and bifurcations. Positive feedbacks occur when the system's own dynamics enhance the effect of a perturbation, leading to an instability. If these positive feedbacks are not controlled via damping mechanisms or negative feedbacks, the system can pass through a "tipping point" into a new domain (Scheffer et al., 2012a). Bifurcations occur when changes in a parameter of the system result in qualitatively different behavior (e.g., stable points become unstable, one stable point becomes multiple stable points). The presence of bifurcations can easily result in abrupt changes. For instance, random fluctuations from within a system (stochastic endogenous fluctuations) can cause the system to depart from an equilibrium or quasi-equilibrium state (e.g., fast, weather time-scale phenomena forcing changes on the longer time-scale climate). Rate-dependent shifts can also occur: a rapid change to an input or parameter of the system may cause it to fail to track changes, and thus tip (Ashwin et al., 2012).

More generally, however, a key characteristic required for abrupt changes to occur is the property of state dependence (aka nonlinearity or nonseperability), where the dynamics (i.e., behavior) of the system are dependent on the system's current state, which may also include its history (time-lagged manifolds). Generically, it is not correct to study these systems using linear methods or by examining variables in isolation (Sugihara et. al. 2012). Overall, research on abrupt changes and tipping points is moving from examining simple systems to investigations of highly connected networks (Scheffer et al., 2012a). The literature on downstream consequences of climate change has not arrived at a clear, common framework analytically as is the case for the physical aspects.

HISTORICAL PERSPECTIVE—PREVIOUS REPORTS ON ABRUPT CHANGE

Here the committee summarizes several previous reports on the topic of abrupt climate change and their recommendations, with the purpose of placing the present study and its recommendations within the context of this previous work. It is particularly instructive to report where progress has been made, and where previous recommendations continue to be echoed but not acted upon.

2002 NRC Report on Abrupt Climate Change

As mentioned above, the first NRC report to comprehensively address abrupt climate change was entitled *Abrupt Climate Change: Inevitable Surprises* (NRC, 2002). This study remains one of the most comprehensive investigations of abrupt climate change to date, addressing the evidence, the potential causes, the potential for the current greenhouse-gas-induced warming to trigger abrupt change, and the potential impacts, ranging from economic to ecological to hydrological to agricultural. One of their key findings was captured in the title of the report, and is summarized by the following quotation:

> "Abrupt climate changes were especially common when the climate system was being forced to change most rapidly. Thus, greenhouse warming and other human alterations of the Earth system may increase the possibility of large, abrupt, and unwelcome regional or global climatic events. The abrupt changes of the past are not fully explained yet, and climate models typically underestimate the size, speed, and extent of those changes. Hence, future abrupt changes cannot be predicted with confidence, and climate surprises are to be expected."

The report made five recommendations in two general categories: implementation of targeted areas of research to expand observations of the present and the past, and implementation of focused modeling efforts.

The first recommendation called for research programs to "collect data to improve understanding of thresholds and non-linearities in geophysical, ecological, and economic systems." The report particularly called out for more work on modes of coupled atmosphere-ocean behavior, oceanic deep-water processes, hydrology, and ice. In the intervening decade, progress has been made in some of these areas. Since 2004 the ocean's meridional overturning circulation has been monitored at 26°N in the North Atlantic (Cunningham et al., 2007). Progress has been made in ice sheet observations from satellites (e.g., Pritchard et al., 2010; Joughin et al., 2010) and in a better understanding of modes of ocean-atmosphere behavior. Nonetheless, as detailed in this report, additional improvements in monitoring for abrupt climate change could be undertaken. For example, the interface between ocean and ice sheet is known to be critical part of ice sheet functioning, yet there are few observations, and no systematic monitoring of the changing conditions at this interface. Also, satellite observations of ice sheets are tenuous as satellites age and funding to replace them, let alone expand their capabilities, is uncertain.

The 2002 report also called for economic and ecological research, a comprehensive land-use census, and development of integrated economic and ecological data sets. Again, some improvements have been made in these areas, notably the National Ecological Observing Network (NEON) to monitor key ecosystem variables in the United States. Other areas, such as a comprehensive land-use census, remain largely unaddressed.

In its second recommendation, the 2002 report called for new, interdisciplinary modeling efforts that would bring together the physical climate system with ecosystems and human systems in an effort to better predict the impacts of abrupt climate change on humans and natural systems, and to better understand the potential for abrupt climate change during warm climate regimes. During the last decade, considerable improvements have been made in many aspects of coupled climate models. Although biases remain, simulation quality has improved for the models in the Coupled Model Intercomparison Project 5 (CMIP5) compared to earlier models (e.g., Knutti et al., 2013). In addition, many climate models have transitioned to include Earth system modeling capabilities (e.g., Hurrell et al., 2013; Collins et al., 2012) in that they incorporate biogeochemical cycles and/or other aspects beyond the standard physical climate model components (atmosphere, ocean, land, sea ice). These new capabilities allow for prognostic simulation of the carbon cycle and the assessment of biogeochemical feedbacks (e.g., Long et al., 2013). In some cases, models now also include the ability to simulate atmospheric chemistry (e.g., Lamarque et al., 2013; Shindell et al., 2013) and large ice sheets (e.g., Lipscomb et al., 2013). This has resulted in more complete system interactions within the model and the ability to investigate additional feedbacks and climate-relevant processes. The inclusion of isotopes into some models (e.g., Sturm et al., 2010; Tindall et al., 2010) is also allowing for a more direct comparison with paleoclimate proxies of relevance to past abrupt change, and a more comprehensive evaluation of the sources and sinks of the atmospheric water cycle that is critical in assessing the risk of future abrupt change and its impacts (e.g., Risi et al., 2010).

Efforts are also underway to more directly link human system interactions into Earth system models. This includes the incorporation of new elements such as agricultural crops (Levis et al., 2012) and urban components (Oleson, 2012). It also includes new efforts to link Integrated Assessment Models to Earth System models (e.g., van Vuuren et al., 2012; Schneider, 1997 Goodess et al., 2003; Bouwman et al., 2006; Warren et al., 2008; Sokolov et al., 2005). Enhancements in model resolution are also enabling the simulation of high-impact weather events of societal relevance (such as tropical cyclones) within climate models (e.g., Jung et al., 2012; Zhao et al., 2009; Bacmeister et al., 2013; Manganello et al., 2012). However, computational resources, while increasing, still remain an obstacle for climate-scale high-resolution simulations. Additionally, model parameterizations and processes need to be reconsidered for simulations at these scales, a task that remains an active research area.

The third recommendation of the 2002 report called for more and better observational data on how our planet and climate system have behaved in the past, with a focus on the high temporal-resolution paleoclimate records required to assess abrupt climate changes. The past decade has witnessed a number of advances in this area, notably terrestrial records from temperate latitudes from cave deposits (e.g., Wang et al., 2008, and references therein), more and better resolved records from ocean sediments (NRC, 2011b), and expanded reconstructions of regional scale hydrological data—including mega-droughts—and changes to the monsoons from records including tree rings and pollen (e.g., Cook et al., 2010c). However, although scientists clearly have an improved understanding of past abrupt climate changes today compared with a decade ago, in many cases the data still remain too sparse spatially to test mechanisms of change using models. Multi-proxy data sets, in which a number of aspects of the climatic and environmental systems are simultaneously reconstructed, remain sparse as well.

The fourth recommendation of the 2002 report focused on improving incorporation of low-probability but high-impact events into societal thinking about climate change. The tendency is to assume a simple distribution of outcomes, and focus on the most probable ones. This approach underestimates the likelihood of extreme events, even ones that would have high impact. If one views risk as the product of likelihood and consequence, then highly consequential, "extreme" events, even if they are unlikely, may pose an equal risk to common events that are not as consequential. The damages resulting from recent extreme weather events (e.g., Hurricane Katrina, Superstorm Sandy, etc.) suggest that there is still a need to better plan for low-probability, high-consequence events, regardless of whether or not their cause is statistically rooted in observed climate trends. That most model predictions for future climate change include more frequent extreme events only heightens the need to take this recommendation seriously.

The fifth and final recommendation of the 2002 report dealt with "no regrets" measures and their application to the potential for abrupt climate change. The report called for taking low-cost steps such as slowing climate change, improving climate forecasts, slowing biodiversity loss, and developing new technologies to increase the adaptability and resiliency of markets, ecological systems, and infrastructure. While there has been some progress in this regard over the past decade, progress has been slow, and remains inadequate to match the scope and scale of the problem. The scientific community has worked to improve climate models, for example, but little has been done to limit greenhouse gas emissions. In fact, the rate of greenhouse gas addition to the atmosphere continues to increase, with many policies in place to accelerate rising greenhouse gases (IMF, 2013). It is sobering to consider that about one-fifth of all fossil fuels ever burned were burned since the 2002 report was released.[2] If indeed, as the 2002 report states, "... greenhouse warming and other human alterations of the Earth system may increase the possibility of large, abrupt, and unwelcome regional or global climatic events", then the danger that existed in 2002 is even higher now, a decade later.

2007 IPCC Fourth Assessment Report

The next major report on climate change following the 2002 report was the 2007 Fourth Assessment Report (AR4) of the Intergovernmental Panel on Climate Change (IPCC, 2007c).[3] The AR4 did not specifically call out abrupt climate change and address it separately, but abrupt climate change was discussed in both the physical science context and in the context of mitigation and adaptation. The Working Group I report, the *Physical Science Basis* (IPCC, 2007b), acknowledged that our understanding of abrupt climate change was notably incomplete and that this limited the ability to model abrupt change, stating that "Mechanisms of onset and evolution of past abrupt climate change and associated climate thresholds are not well understood. This limits confidence in the ability of climate models to simulate realistic abrupt change." However, the Working Group I report did specifically address the issue of a shutdown of the formation of North Atlantic Deep Water and concluded from modeling studies that although it was very likely (>90 percent chance) that the deep water formation would slow in the coming century, it was very unlikely (<10 percent chance) that this process would undergo a large abrupt transition, at least in the coming decades to century. This was an important advancement in the understanding of the potential threats of

[2] Sum of global emissions from 1751 through 2009 inclusive is 355,676 million metric tons of carbon; sum of global emissions from 2002 through 2009 inclusive is 64,788 million metric tons of carbon (Boden et al., 2011). Total carbon emissions for 2002-2009 compared to the total 1751-2009 is thus greater than 18%.

[3] The Working Group I report of the Fifth Assessment Report (AR5) of the IPCC was released after this report had been submitted for peer-review. The Committee drew their conclusions from the broader scientific literature, which is also the basis for IPCC AR5. Although this report only references the IPCC AR5 in a few instances, the broader conclusions of this report are consistent with the IPCC AR5.

abrupt climate change, and an example of a threat that has been categorized as less likely due to improved understanding of the process.

The AR4 Working Group 2 (WG2) report, *Impacts, Adaptation and Vulnerability* (IPCC, 2007a), addresses abrupt climate change throughout the report, and summarizes the impacts of extreme events and key vulnerabilities including topics such as coastal inundation, food supply disruption, and drought. The AR4 WG2 report repeatedly calls for more research to be done on the impacts of abrupt change, particularly a collapse of the North Atlantic Deep Water formation (which was not considered likely) and a relatively rapid sea level rise of many meters due to rapid (century-scale) loss of ice from Greenland and/or West Antarctica, noting that without a better scientific understanding of the potential impacts, it was impossible to carry out impact assessments. That report also notes that there has been "little advance" on the topic of "proximity to thresholds and tipping points."

The AR4 Working Group 3 report on *Mitigation of Climate Change* (Metz et al., 2007) mentions abrupt climate change, but does not consider the topic in detail. It acknowledges that abrupt climate changes are not well incorporated into conventional decision-making analysis, which tends to enable substantial vulnerability to high-impact, low-probability events. This potentially increases the damages from any such events that could occur—and perhaps even the probability of such events—through lack of mitigation and adaptation. Similarly, abrupt climate change can challenge assumptions made in economic cost-benefit analyses, for example the cost of a lost species versus the savings realized in not acting to save that species.

2008 USCCSP Synthesis and Assessment Product 3.4: Abrupt Climate Change

The next major report to address abrupt climate change was the 2008 United States Climate Change Strategic Plan *Synthesis and Assessment Product 3.4, Abrupt Climate Change* report (USCCSP, 2008). This report (also known as SAP 3.4) was focused solely on abrupt climate change, but took a different approach from the 2002 NRC report by focusing on four key areas of interest:

1. Rapid Changes in Glaciers and Ice Sheets and their Impacts on Sea Level;
2. Hydrological Variability and Change;
3. Potential for Abrupt Change in the Atlantic Meridional Overturning Circulation (AMOC); and
4. Potential for Abrupt Changes in Atmospheric Methane.

As stated in their introduction,"This SAP picks up where the NRC report and the IPCC AR4 leave off, updating the state and strength of existing knowledge, both from the paleoclimate and historical records, as well as from model predictions for future change." Their findings are woven into the present report, but are too extensive to repeat in this Introduction. A few key findings are discussed briefly, however.

"Although no ice-sheet model is currently capable of capturing the glacier speedups in Antarctica or Greenland that have been observed over the last decade, including these processes in models will very likely show that IPCC AR4 projected sea level rises for the end of the 21st century are too low." This finding re-states the caveat expressed in the AR4 concerning the lack of understanding about glacial dynamics, particularly fast-flowing, large glaciers such as parts of Greenland and West Antarctica. As detailed in the present report, the scientific community has not yet formed a consensus regarding the rate with which large glaciers can shed ice, and thus uncertainty remains about the speed and eventual magnitude of sea level rise, both over this coming century, and beyond.

The SAP 3.4 raised two questions concerning tipping points in droughts. The first is the model predicted expansion of aridity into the U.S. Southwest accompanying the general warming of the ocean and atmosphere. As they state, "If the model results are correct, then this drying may have already begun, but currently cannot be definitively identified amidst the considerable natural variability of hydroclimate in Southwestern North America." This remains a key area of concern, and one that is addressed in this report. The SAP 3.4 also raised the issue of monitoring for tipping point behavior in the hydrological cycle (Chapter 4 of that report), including the potential for mega-droughts in a world warmed by greenhouse gases. Physical understanding suggests that mega-droughts are more likely to be triggered by interior reorganization of the ocean-atmosphere system rather than by overall warming of Earth's surface, although overall warming can cause interior reorganization and thus can be responsible indirectly. The SAP 3.4 report states that it is unclear whether current climate models are capable of predicting the onset of mega-droughts: "... systematic biases within current coupled atmosphere-ocean models raise concerns as to whether they correctly represent the response of the tropical climate system to radiative forcing and whether greenhouse forcing will actually induce El Niño/Southern Oscillation-like patterns of tropical SST change that will create impacts on global hydroclimate...". Research done since SAP 3.4 suggests that the drying from human-caused climate change (radiatively forced reduction of the net surface water flux, i.e., the precipitation minus evapotranspiration) appears to be comparable to the drying induced by the impacts of La Nina over the Southwestern North America since 1979 (Seager and Naik, 2012). In the future, drying forced by the addition of anthropogenic greenhouse gases to the atmosphere is expected to increase along with earlier melting and reduced storage of mountain snow packs, although whether changes of climate variability would intensify or mitigate such drying remains uncertain (Seager and Vecchi, 2010). In addition, as increasing anthropogenic forcing shifts the surface temperature distribution (Trenberth et al., 2007; Meehl et al., 2007b), extreme warm temperatures and soil moisture loss would increase. Thus, the "climate dice", mainly controlled by random climate variability, would become more "loaded" with the risk of mega-drought even if a particular drought is simply the result of natural climate variability (Hansen et al., 2012, 2013a).

The SAP 3.4 also addressed the potential for tipping points in the North Atlantic Deep Water formation; and in the release of methane, a potent greenhouse gas, to the atmosphere. As with the IPCC AR4 report, the SAP 3.4 report concluded that deep-water formation was not likely to "tip," although it is likely to decrease, with impacts on precipitation patterns that could be tipped on regional scales. The potential for catastrophic methane release, from decomposition of terrestrial carbon stocks in permafrost, or methane ice in clathrates, was considered small. However, the potential for gradually increasing methane and CO_2 release from thawing permafrost was considered important, and would accelerate the loading of these greenhouse gases into the atmosphere over many decades to centuries. The report recommended that the United States should "Prioritize the monitoring of atmospheric methane abundance and its isotopic composition with spatial density sufficient to allow detection of any change in net emissions from northern and tropical wetland regions." Such a prioritization has not occurred; in fact, the primary monitoring network for greenhouse gases globally, the NOAA network,[4] has faced funding cuts of over 30 percent in the past several years.

2012 NRC Report on Climate and Social Stress

The NRC report *Climate and Social Stress: Implications for Security Analysis* (2012b) is the most recent report to address abrupt climate change. It dedicates a section to a general discussion of abrupt climate change, with an additional section allocated to the topic of extreme events. The report focuses on the coming decade, and as such they conclude that there is little expectation in the scientific community for an abrupt change on that timescale. It makes several recommendations including enhanced monitoring, such as enhanced drought metrics to assess if a region is entering a new mega-drought. These include social factors as well, for example:

[4] http://www.esrl.noaa.gov/gmd/ccgg/flask.html.

> "changes in the social, economic, and political factors that affect the size of the exposed populations, their susceptibility to harm, the ability of the populations to cope, and the ability of their governments to respond. Where potentially affected areas are important producers of key global commodities such as food grains, it would also be important to assess the effects of climate-induced supply reductions on global markets and vulnerable populations."

The NRC 2012 report also called for enhanced monitoring of such factors, and noted that society is, in general, rather blind to what is at risk to abrupt climate change, for example, having only limited understanding about the risks posed by sea level rise to coastal infrastructure, toxic materials in landfills, or drinking water aquifers.

THIS REPORT

Looking back across these previous reports, it can be seen that while a great deal of progress on the topic of abrupt climate change has been made, there is still a long way to go to achieve an understanding of these issues with enough fidelity to be able to anticipate their occurrence. This report takes on that challenge, as per the committee's statement of task, given in Box 1.5.

Organization of the Report

The committee recognized that discussions of abrupt climate changes and impacts of abrupt climate changes may have different audiences. As such, the report is organized so that one can seek information on the processes as well as information on the impacts.

Chapter 2 gives examples of abrupt climate changes, specifically those examples that the committee believes are worthy of highlighting either because they are currently believed to be the most likely and the most impactful, because they are predicted to potentially cause severe impacts but with uncertain likelihood, or because they are now considered to be unlikely to occur but have been widely discussed in the literature or media. This section includes processes such as the changing chemistry of the oceans and the melting of ice sheets leading to sea level rise. Many of these processes have been discussed in the recent reports (Box 1.2), and the committee provides an updated discussion building on those previous reports.

Chapter 3 discusses abrupt climate impacts from the perspective of how they affect humans, building on many of the same processes discussed in Chapter 2. Examples include abrupt changes in food availability, water availability, and ecosystem services.

BOX 1.5 STATEMENT OF TASK

This study will address the likelihood of various physical components of the Earth system to undergo major and rapid changes (i.e., abrupt climate change) and, as time allows, examine some of the most important potential associated impacts and risks. This study will explore how to monitor climate change for warnings of abrupt changes and emerging impacts. The study will summarize the current state of scientific understanding on questions such as:

1. What is known about the likelihood and timing of abrupt changes in the climate system over decadal timescales? Are any of the phenomena considered by the committee currently embodied in computational climate models? The committee could consider relevant physical and biological phenomena such as:
   - large, abrupt changes in ocean circulation and regional climate;
   - reduced ice in the Arctic Ocean and permafrost regions;
   - large-scale clathrate release;
   - changes in ice sheets;
   - large, rapid global sea-level rise;
   - growing frequency and length of heat waves and droughts;
   - effects on biological systems of permafrost/ground thawing (carbon cycle effects);
   - phase changes such as cloud formation processes; and
   - changes in weather patterns, such as changes in snowpack, increased frequency and magnitude of heavy rainfall events and floods, or changes in monsoon patterns and modes of interannual or decadal variability.

2. For the abrupt climate changes and resulting impacts identified by the committee, what are the prospects for developing an early warning system and at what lead time scales? What can be monitored to provide such warnings? What monitoring capabilities are already in place? The committee will consider monitoring capabilities that include both direct observations and the use of models in conjunction with observations.

3. What are the gaps in our scientific understanding and current monitoring capabilities? What are the highest priority needs for future research directions and monitoring capabilities to fill those gaps?

Chapter 4 examines the way forward in terms of both research on abrupt changes and their impacts, and monitoring to detect and potentially predict abrupt changes. This chapter examines priorities and capabilities for addressing research knowledge gaps. It also addresses the question of what to monitor to observe that an abrupt change is coming, and how to identify tipping points in various systems.

CHAPTER TWO

Abrupt Changes of Primary Concern

The following section describes potential abrupt climate changes that are of primary concern, either because they are currently believed to be the most likely and the most impactful, because they are predicted to potentially cause severe impacts but with uncertain likelihood, or because they are considered to be unlikely to occur but have been widely discussed in the literature or media. As such, the Committee did not attempt to create a comprehensive catalog of potential abrupt changes. As described in the Introduction, this section examines both abrupt climate changes in the physical climate system itself and abrupt climate impacts in physical, biological, or human systems that are triggered by a steadily changing climate.

ABRUPT CHANGES IN THE OCEAN

The Atlantic Meridional Overturning Circulation

The Atlantic Meridional Overturning Circulation (AMOC)—characterized by warm surface waters flowing northward and cold deep waters flowing southward throughout the Atlantic basin—is defined as the zonal integral of the northward mass flux at a particular latitude. The deep limb of this overturning circulation carries waters that are formed via convection in the Nordic and Labrador Seas (Figure 2.1). Collectively, these waters constitute North Atlantic Deep Water, which is exported to the global ocean at depths between about 1000 and 4000 m. The southward-flowing deep limb of the overturning circulation is compensated by an upper limb of northward-flowing surface waters, which head to the Nordic and Labrador Seas to replenish the regions of convection. Together, the upper and lower limbs of the overturning circulation produce a poleward flux of heat that has strong global and regional impacts. The AMOC also plays an important role in the transport of carbon in the Atlantic. Thus, variability in the AMOC's strength is of much interest, as a diminishment or strengthening would impact the ocean's effectiveness as a heat and carbon reservoir.

Examinations of paleoclimate temperatures and other variables recorded in both North Atlantic ocean sediments and Greenland ice cores (e.g., Lehman and Keigwin, 1992; Alley et al., 1993; Taylor et al., 1993) have led to suggestions that the AMOC

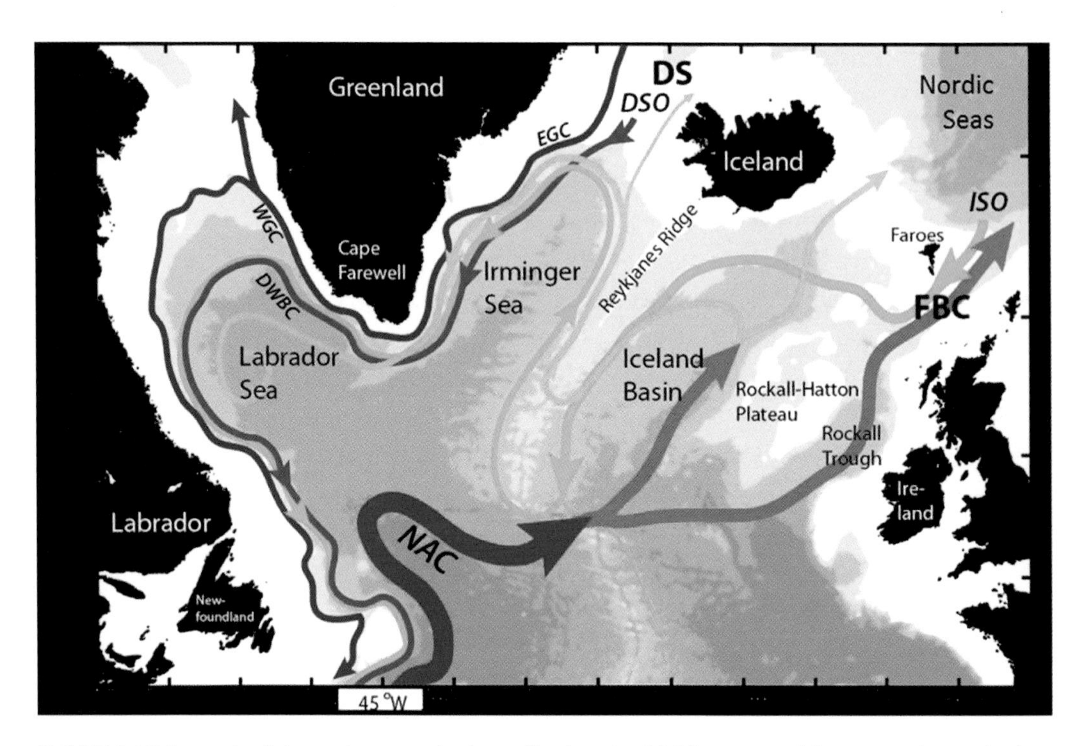

FIGURE 2.1 Schematic of the major warm (red to yellow) and cold (blue to purple) water pathways in the North Atlantic subpolar gyre. Acronyms not in the text: Denmark Strait (DS); Faroe Bank Channel (FBC); East and West Greenland Currents (EGC, WGC); North Atlantic Current (NAC); DSO (Denmark Straits Overflow); ISO (Iceland-Scotland Overflow). Figure courtesy of H. Furey (WHOI).

abruptly changed in the past. Following on this examination, questions have arisen as to the possible likelihood of an abrupt change in the future.

The Stability of the Atlantic Meridional Overturning Circulation

Climate and Earth system models are used to understand potential changes in the AMOC, including potential feedbacks in the system, although the representation of unresolved physics (such as the parameterization of ocean mixing) could potentially be of concern in long, centennial simulations. Because saltier water is denser and thus more likely to sink, the transport of salt poleward into the North Atlantic provides a potentially destabilizing advective feedback to the AMOC (Stommel, 1961); i.e., a reduction in the strength of the AMOC would lead to less salt being transported into the North Atlantic, and hence a further reduction in the AMOC would ensue. As noted by Rahmstorf (1996), the presence of this slow salt-advection feedback is critical to the existence of stable multiple equilibria.

Climate and Earth system models have been used to investigate the stability of the AMOC, in particular the number of stable states that the system can exist in, which is an important characteristic to know for fully understanding the climate system. Carefully designed non-linear modeling experiments using Earth system Models of Intermediate Complexity (EMICs; and also the FAMOUS AOGCM; Hawkins et al., 2011) have revealed a model-dependent threshold beyond which an active AMOC cannot be sustained (Rahmstorf et al., 2005; see Figure 2.2). However, analysis of the AMOC in the models that submitted simulations in support of the third phase of the Community Model Intercomparison Project[1] (CMIP3; Meehl et al., 2007a) suggested that the CMIP3 models were overly stable (Drijfhout et al., 2011; Hofmann and Rahmstorf, 2009), i.e., that an abrupt change in the AMOC was not likely to be simulated in the models even if it were to be likely in reality.

Several studies (de Vries and Weber, 2005; Dijkstra, 2007; Weber et al., 2007; Huisman et al., 2010; Drijfhout et al., 2011; and Hawkins et al., 2011) have suggested that the sign of the net freshwater flux into the Atlantic across its southern boundary via the overturning circulation determines whether or not the AMOC is in a monostable or bistable regime. Observations suggest that the present day ocean resides in a bistable regime, thereby allowing for multiple equilibria and a stable "off" state of the AMOC (Hawkins et al., 2011). By examining the preindustrial control climate of the CMIP3 models, Drijfhout et al. (2011) found that the salt flux was mostly negative (implying a positive freshwater flux), indicating that these models were mostly in a monostable regime. This was not the case in the CMIP5 models where Weaver et al. (2012) found that 40 percent of the models were in a bistable regime throughout their integrations. Although this question of the number of stable states of the system is important for a complete understanding of the climate system, it is important to emphasize that regardless of this stability question, the CMIP5 models also show no evidence of an abrupt collapse for the 21st century.

In addition to the main threshold for a complete breakdown of the circulation, other thresholds may exist that involve more-limited changes, such as a cessation or diminishment of Labrador Sea deep water formation (Wood et al., 1999). Rapid melting of the Greenland ice sheet causes increases in freshwater runoff, potentially weakening the AMOC. None of the CMIP5 simulations include an interactive ice sheet component. However, Jungclaus et al. (2006), with parameterized freshwater melt as high

[1] http://www-pcmdi.llnl.gov/ipcc/about_ipcc.php.

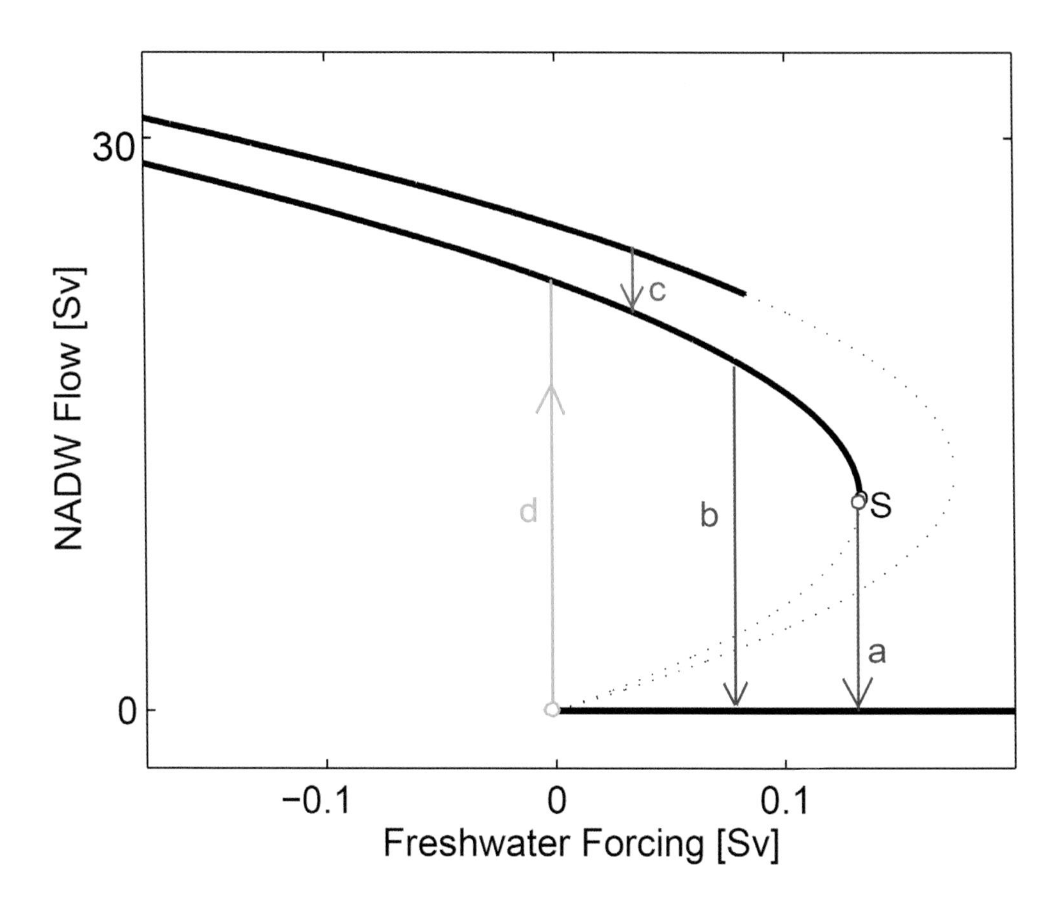

FIGURE 2.2 Schematic diagram illustrating the hysteresis behavior of the equilibrium strength of the AMOC in response to the addition of a North Atlantic surface freshwater perturbation of variable magnitude. Positive values indicate the sustained addition of freshwater to the surface; negative values indicate the sustained subtraction of freshwater from the surface; the zero value corresponds to the present-day situation. The two upper heavy branches indicate the possibility of multiple states with different convection sites. Transitions between stable equilibria of the AMOC with and without active deepwater formation are indicated by: (a) transition associated with slow advective instability, (b) transition associated with fast convective instability, and (d) initiation of convection and subsequent spin-up of North Atlantic Deep Water (NADW) formation. The S indicates the point beyond which a stable equilibrium with active NADW formation cannot exist. (c) indicates a possible transition between active modes of NADW formation with different location of convection.

Note: Hysteresis is defined as "a lag in response exhibited by a body in reacting to changes in forces" (Random House Kernerman Webster's College Dictionary) and is used in many fields such as engineering, economics, biology, etc. to refer to a system that depends on the current but also past environmental conditions.

SOURCE: Rahmstorf, 1999.

as 0.09 Sv, and Hu et al. (2009), using the NCAR Community Climate System Model (CCSM) with year 2000 initial parameterized freshwater melt of 0.01 Sv increasing at a rate of 1 percent/year, 3 percent/year and 7 percent/year, as well as Mikolajewicz et al. (2007) and Driesschaert et al. (2007), using coupled ice-sheet/climate models, found only a slight temporary effect of increased melt water fluxes on the AMOC. The impact of these fluxes on the AMOC was generally small compared to the effect of enhanced poleward atmospheric moisture transport and ocean surface warming; or it was only noticeable in the most extreme scenarios. But this point needs to be further quantified.

While many more model simulations were conducted in support of the IPCC AR5 (Collins et al., 2012) under a wide range of forcing scenarios, projections of the behavior of the AMOC over the 21st century and beyond have changed little from what was reported in the IPCC AR4 (Meehl et al., 2007b). In the case of the CMIP5 models, Weaver et al. (2012) showed that the behavior of the AMOC was similar over the 21st century under four very different radiative forcing scenarios (RCP 2.6; RCP4.5; RCP 6.0; RCP8.5—these Representative Concentration Pathways [RCPs] are detailed in Moss et al., 2010). All models found a 21st century weakening of the AMOC with a multi-model average of 22 percent for RCP2.6, 26 percent for RCP4.5, 29 percent for RCP6.0 and 40 percent for RCP8.5. While two of the models eventually realized a slow shutdown of the AMOC under RCP8.5 (the scenario with the largest amount of warming), none exhibited an abrupt change of the AMOC.

The similarity of the model responses despite the widely varying transports of salt into the North Atlantic across its southern boundary (and hence sign and magnitude of the salt advection feedback) suggests that like the CMIP3 models (Gregory et al., 2005), the reduction of the AMOC in the global warming experiments performed by the CMIP5 models is mainly driven by local changes in surface thermal flux rather than surface freshwater flux. North Atlantic surface warming decreases water density there, thus reducing the rate of sinking. In addition, as noted above, none of the CMIP models incorporated the additional freshwater effects of ice sheet melting. This is an important caveat since asymmetric freshwater forcing is capable of initiating a fast, convective instability that could cause the AMOC to abruptly shut down if it were in a bistable regime and suitably close to its stability threshold. This would explain why abrupt changes of the AMOC appear to be pervasive features of the paleoclimate record when vast reservoirs of freshwater were available in the form of ice and proglacial lakes on land.

A question that needs to be further addressed is the extent to which projected changes in Greenland ice sheet melting could affect the amount and location of freshwater release into the North Atlantic and hence the subsequent evolution of the AMOC. As noted in Meehl et al. (2007b) it is very unlikely that the AMOC will undergo an abrupt transition or collapse in the 21st century. Delworth et al. (2008) pointed out that for an abrupt transition of the AMOC to occur, the sensitivity of the AMOC to forcing would have to be far greater than that seen in current models. Alternatively, significant ablation of the Greenland ice sheet greatly exceeding even the most aggressive of current projections would be required. As noted in the ice sheet section later in this chapter, Greenland ice has about 7.3m equivalent of sea level rise, which, if melted over 1000 years, yields an annual rise rate of 7 mm/yr, about 2 times faster just from Greenland than today's rate from all sources, and more than 10 times faster than the rate from Greenland over 2000–2011 (Shepherd et al., 2012). Although neither possibility can be excluded entirely, it is unlikely that the AMOC will collapse before the end of the 21st century because of global warming.

Observations of the Atlantic Meridional Overturning Circulation

Recent observational studies have focused on ascertaining two questions of relevance to the AMOC response to climate change: What is the impact of variable North Atlantic Deep Water production on the ocean's meridional overturning? And, what is the current state of the AMOC and its variability? Studies relevant to both questions are briefly reviewed here (material drawn from Lozier, 2012).

Though many modeling studies have demonstrated the impact of deep water formation changes on the overturning circulation, the observational evidence for such a linkage has been hard to come by for two reasons: (1) Deep water formation is difficult to quantify because the time and locale of production are highly variable from winter to winter, and (2) overturning circulation measures require observations that span the basin, which have been limited in space and time. Because of this second difficulty, a measure of the Deep Western Boundary Current (DWBC) transport has traditionally been considered a shortcut to the measure of the AMOC: while the upper limb of the AMOC was considered inextricably linked to the much more energetic wind-driven circulation, the lower limb was considered to be "channeled" through the DWBC.

An opportunity to assess the linkage between deep water formation variability and DWBC changes was afforded by the deployment of a moored array east of the Grand Banks (Clarke et al., 1998; Meinen et al., 2000; Schott et al., 2006). In an extensive analysis of the time series from these two deployments, Schott et al. (2006) found that the transport rates of Labrador Sea Water (LSW) over these two time periods were remarkably similar despite the large differences in convective activity in the Labrador Sea during the two time periods: the earlier time period was marked by strong convective activity, while LSW production was considerably weaker during the latter time period (Lazier et al., 2002). This result raised questions about the responsiveness of the AMOC to changes in deep water production; however, the linkage could not be conclusively ruled out because of increasing indications that the DWBC was not the sole conduit for the passage of deep waters to the lower latitudes (Schott et al., 2006). And in fact, recent observational (Lavender et al., 2000; Fischer and Schott, 2002; Bower et al., 2009) and modeling studies (Gary et al., 2011; Lozier et al., 2010) of subsurface floats have revealed that the DWBC is not the sole, and perhaps not even the dominant, conduit for the transport of the waters within the deep limb of the AMOC. Thus, a measure of the DWBC is no longer considered a sufficient monitor of AMOC changes.

For a full accounting of the AMOC and its variability, it is now understood that trans-basin measurements of transport are necessary. Attempts to understand trans-basin AMOC variability over the modern observational record traditionally have had to rely on indirect estimates assessed from hydrography. Bryden et al. (2005) used five repeat surveys at 25°N from 1957 to 2004 to show that the overturning slowed by 30 percent over the period of the surveys, an astounding and unanticipated change over such a relatively short time. However, an assessment of transports at 48°N using five repeat World Ocean Circulation Experiment sections and air-sea heat and freshwater fluxes as input to an inverse box model yielded no significant trend in the meridional overturning at that latitude (Lumpkin et al., 2008), though the time period studied was relatively short (1993-2000).

In 2004 an observational system was put in place to provide the first continuous measure of the AMOC (Cunningham et al., 2007). The RAPID/MOCHA program (Rapid Climate Change/Meridional Overturning Circulation and Heatflux Array) comprises instruments deployed along a section at 25°N stretching from the North American continent to the west coast of Africa. After just one year of measurements, the conceptual understanding of overturning variability changed dramatically. As seen in Figure 2.3, the overturning strength changed six-fold from April of 2004 to April of 2005, from a minimum of ~5 Sv to a maximum of ~30 Sv. With the demonstrated intraseasonal variability, synoptic sections were now understood to be inadequate to capture measures of interannual transport variability. The continuation of the time series has revealed a strong seasonality (Rayner et al., 2011) that dominates the record, as well as strong intrannual variability (McCarthy et al., 2012).

Unfortunately, the strong intraseasonal variability of the AMOC revealed by the RAPID/MOCHA array seriously constrains our ability to recreate AMOC variability over the modern observational period, since synoptic hydrographic sections are the

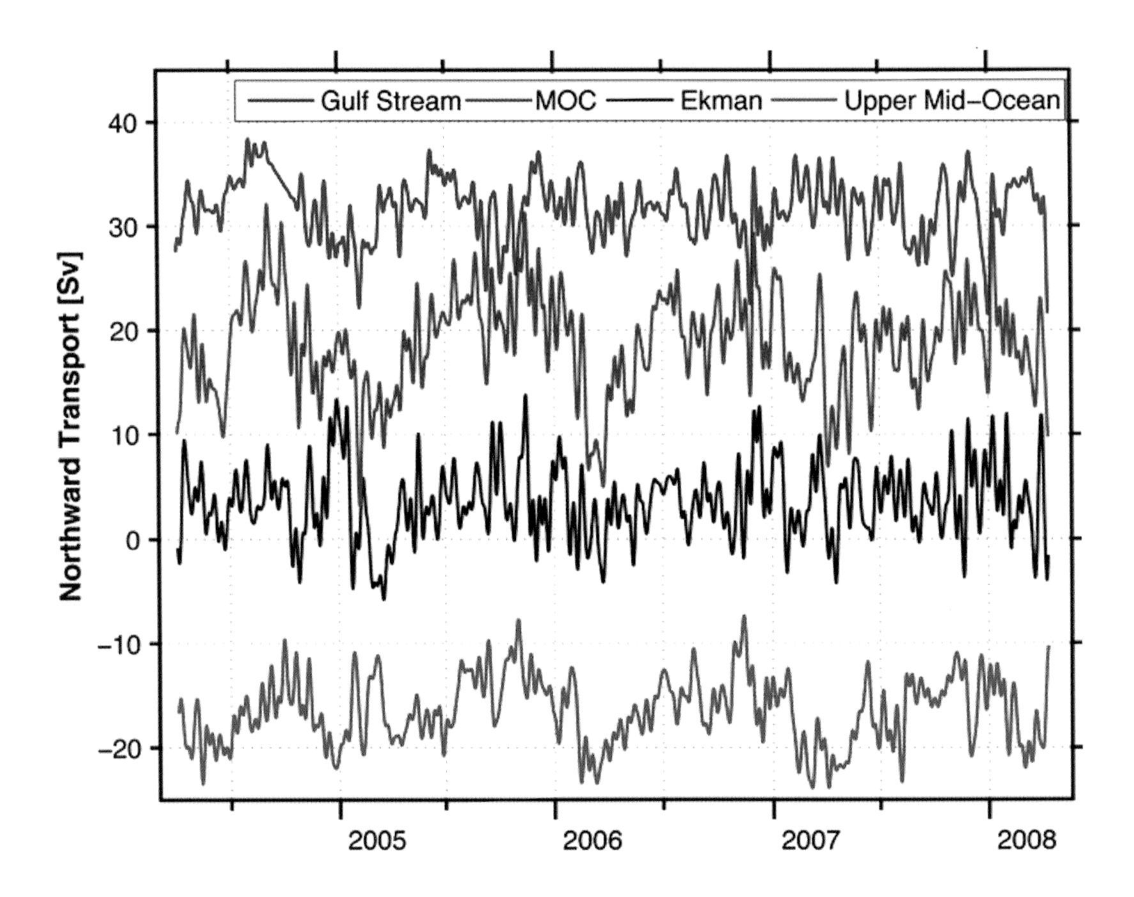

FIGURE 2.3 Time series of the meridional overturning transport at 26° N from the RAPID/MOCHA array. The meridional overturning circulation (MOC) is computed from the sum of the Gulf Stream transport through the Florida Straits, directly measured via electromagnetic cables; the Ekman transport, estimated from QuikSCAT winds; and the midocean geostrophic transport, estimated from the moored array instruments. Importantly, this time series demonstrates the significant interannual transport variability. SOURCE: Rayner et al., 2011.

only past trans-basin measurements. Furthermore, as detailed in a recent review by Cunningham and Marsh (2010), modeling estimates have been unable to help in this regard: there is currently no consensus on the strength of the AMOC in assimilation/re-analysis products, and ocean general circulation models are in disagreement about the strength and variability of the AMOC. Indeed, an active area of research within the climate modeling community is focused on the cause for such wide ranges of AMOC estimates from state estimates that are drawing from the same observational databases (U.S. CLIVAR Project Office, 2011) and in ocean simulations forced with the same atmospheric conditions (e.g., Danabasoglu et al., 2013). It is important to note that the models run for the IPCC discussed above also have varying AMOC strength and interannual variability, yet they are in agreement on a lack of abrupt change for the 21st century.

In lieu of consistent model estimates, proxy measures of the overturning derived from satellite altimetry and Argo float data are appealing, but to date these measures have been limited to latitudes of steep topography on the western boundary (Willis, 2010), and are of limited duration to provide a temporal context of decades. Thus, to date direct AMOC observations are limited to one latitude (26°N), and past measures of change remain elusive. Although the RAPID array is providing unprecedented measurements, recent modeling and data analysis studies (Bingham et al., 2007; Baehr et al., 2009; Lozier et al., 2010; Biastoch et al., 2008a; Biastoch et al., 2008b) reveal gyre-specific measures of the AMOC, suggesting that the AMOC variability measured by the RAPID array cannot safely be assumed representative of AMOC variability outside of the North Atlantic subtropical basin.

Summary and the Way Forward

Although models do not indicate that AMOC is likely to change abruptly in the coming decades, it is important to monitor the North Atlantic to confirm the understanding of how AMOC responds to a changing climate. Observational studies over the past decade or so reveal a meridional overturning circulation with a tenuous link to the production of deep water masses via local overturning at high latitudes in the North Atlantic. However, the deep ocean remains vastly undersampled, particularly so with respect to measures appropriate for the calculation of AMOC variability. To ascertain with confidence the extent to which deep water production impacts the ocean's meridional circulation and hence the ocean's contributions to the global poleward heat flux, continuous measures of trans-basin mass and heat transports are needed. Although such measurements are underway with the RAPID/MOCHA array, the studies cited above have made it increasingly clear that AMOC fluctuations are coherent over only limited meridional distances: break points in coherence occur at key latitudes, in particular at the subpolar/subtropical gyre boundary in the North Atlantic. Therefore, a transoceanic line in the subpolar North Atlantic, currently being planned by the international community, that measures the net contributions of the overflow waters from the Nordic Seas as well as those from the Labrador Sea, to the AMOC, would directly test the legitimacy of the decades-long supposition that variability in North Atlantic Deep Water production translates into meridional overturning variability (Figure 2.4). This measurement system would—in conjunction with the RAPID/

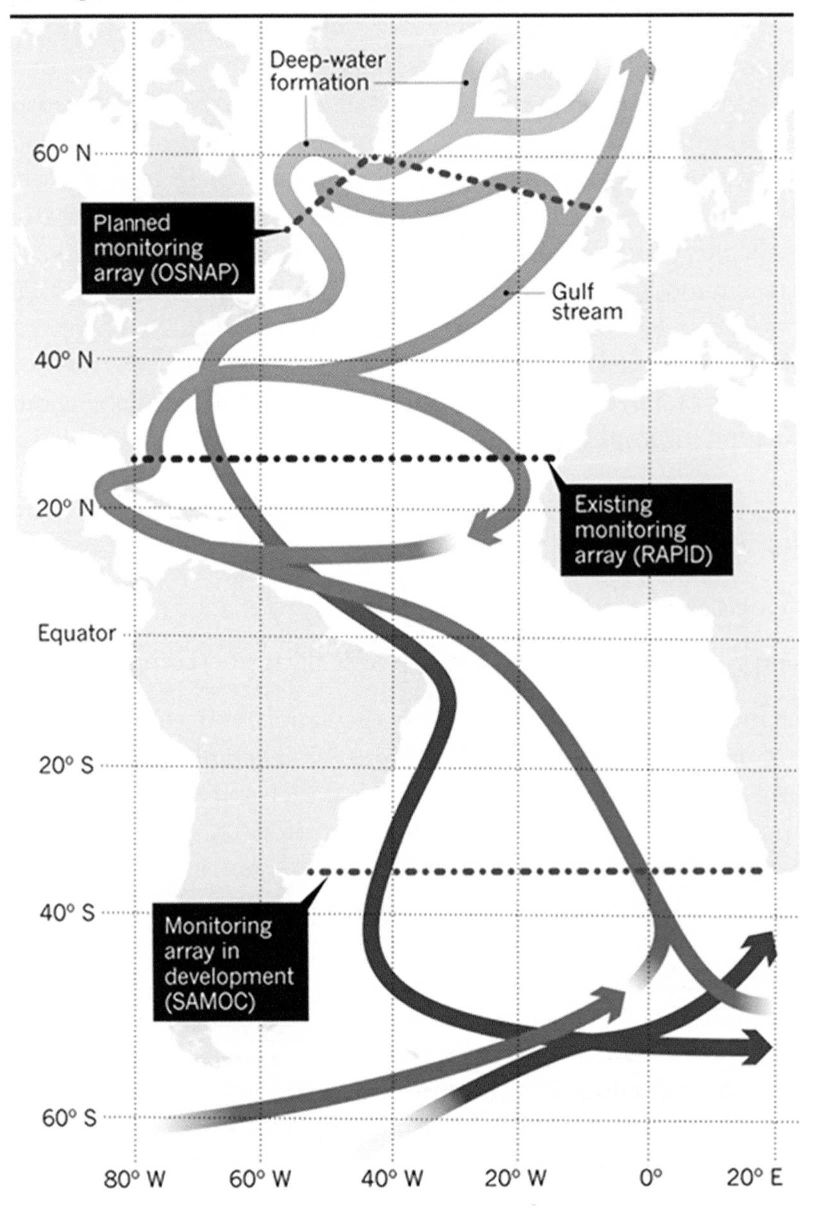

FIGURE 2.4 Existing and proposed monitoring locations for the Atlantic Ocean. Source: Adapted from Schiermeier, 2013.

MOCHA array—provide a means to evaluate intergyre connectivity within the North Atlantic and allow for a determination of how and whether deep water mass formation impacts overturning and poleward heat and freshwater transports throughout the North Atlantic. Additionally, such an observing system, by measuring the temporal and spatial variability of the AMOC for approximately a decade, would provide essential ground truth to AMOC model estimates and would also yield insight into whether AMOC changes or other atmospheric/oceanic variability have the dominant impact on interannual sea surface temperature (SST) variability. To make clear assessments of the AMOC's response to anthropogenic climate change, it is expected that a multi-decadal observing system will be necessary. An observing system serving this purpose would be one where a few critical in situ observations, coupled with satellite observations and the Argo float array, provide a reliable and sustainable measure of the AMOC for decades to come.

Ice Sheets and Sea Level

Based on both simple physics and observations of the past, there is high confidence in the conclusion that sea level rises in response to warming. Sea-level rise can have large impacts (e.g., Nicholls et al., 2007), such as damage to or loss of infrastructure near coasts, loss of freshwater supplies, and displacement of people whose homes are lost to a rising ocean. Although sea-level rise typically is slow compared to many environmental changes, even this type of gradual sea-level rise may force other systems to cross thresholds and trigger abrupt impacts for natural or human systems unless adaptive measures are taken. For example, rising sea level increases the likelihood that a storm surge will overtop a levee or damage other coastal infrastructure, such as coastal roads, sewage treatment plants, or gas lines—all with potentially large, expensive, and immediate consequences (Nordhaus, 2010). (See Box 2.1 for discussion of vulnerabilities of US coastal infrastructure.)

A separate but key question is whether sea-level rise itself can be large, rapid and widespread. In this regard, rate of change is assessed relative to the rate of societal adaptation. Available scientific understanding does not answer this question fully, but observations and modeling studies do show that a much faster sea-level rise than that observed recently (~3 mm/yr over recent decades) is possible (Cronin, 2012). Rates peaked more than 10 times faster in Meltwater Pulse 1A during the warming from the most recent ice age, a time with more ice on the planet to contribute to the sea-level rise, but slower forcing than the human-caused rise in CO_2 (Figure 2.5 and 2.6). One could term a rise "rapid" if the response or adaptation time is significantly longer than the rise time. For example, a rise rate of 15 mm/yr (within the range of projec-

BOX 2.1 VULNERABILITY OF U.S. COASTAL INFRASTRUCTURE

Thirty nine percent of the population lives in coastal shoreline counties. This population grew by 39 percent between 1970 and 2010, and is projected to grow by 8.3 percent by 2020. The population density of coastal counties is 446 people per sq mile, which is over 4 times that of inland counties.

Just under half of the annual GDP of the United States is generated in coastal shoreline counties, an annual contribution that was $6.6 trillion in 2011. If counted as their own country, these counties would rank as the world's third largest economy, after the United States and China. Some portions of these counties are well above sea level and not vulnerable to flooding (e.g., Cadillac Mountain, Maine, in Acadia National Park, at 470 m). But, the interconnected nature of roads and other infrastructure within political divisions mean that sea-level rise would cause problems even for the higher parts of these counties. The following statistics, from NOAA's State of the Coast,[a] highlight the wealth and infrastructure at risk from rising seas:

- **$6.6 trillion:** Contribution to GDP of the coastal shoreline counties, just under half of US GDP in 2011.[b]
- **51 million:** Total number of jobs in the coastal shoreline counties of the US in 2011.[c]
- **$2.8 trillion:** Wages paid out to employees working at establishments in the coastal shoreline counties in 2011.[d]
- **3:** Global GDP rank (behind the United States and China) of the coastal shoreline counties, if considered an individual country.[e]
- **39%:** Percent of the nation's total population that lived in coastal watershed counties in 2010 (less than 20 percent of the total land area excluding Alaska).[f]
- **34.8 million:** Increase in US coastal watershed county population from 1970 to 2010 (or a 39 percent increase).[g]
- **446 persons/mi^2:** Average population density of the coastal watershed counties (excluding Alaska). Inland density averages 61 persons per square mile.[h]
- **37 persons/mi^2:** Expected increase in US coastal watershed county population by 2020 (or an 8.3 percent increase).[i]

Projections of sea-level rise remain notably uncertain even if the increase in greenhouse gases is specified accurately, but many recently published estimates include within their range of possibilities a rise of 1m by the end of this century (reviewed by Moore et al., 2013). For low-lying metropolitan areas, such as Miami and San Francisco, such a rise could lead to significant flooding (Figure A) (NRC, 2012e; Strauss et al., 2012; Tebaldi et al., 2012). In many cases, such areas would be difficult to defend by dikes and dams, and such a large sea level rise would require responses ranging from potentially large and expensive engineering projects to partial or near-complete abandonment of now-valuable areas as critical infrastructure such as sewer systems, gas lines, and roads are disrupted, perhaps crossing tipping points for adaptation (Kwadijk et al., 2010). Miami was founded little more than one century ago, and could face the possibility of sea level rise high enough to potentially threaten the city's critical infrastructure in another century (Strauss et al., 2013). In terms of modern expectations for the lifetime of a city's infrastructure, this is abrupt. If sometime in the coming centuries sea level should rise 20 to 25 m, as suggested

BOX 2.1 CONTINUED

FIGURE A Elevation map of Miami, Florida. The low elevation of many parts of the city and surroundings, combined with a water-permeable sand and coral base, make it particularly vulnerable to sea-level rise. Areas at risk from a 1-meter rise in sea level are shown, where 1 meter is within the range of many recently published estimates for sea-level rise by the end of this century. SOURCE: Cool Air Clean Planet, http://cleanair-coolplanet.org/.

BOX 2.1 CONTINUED

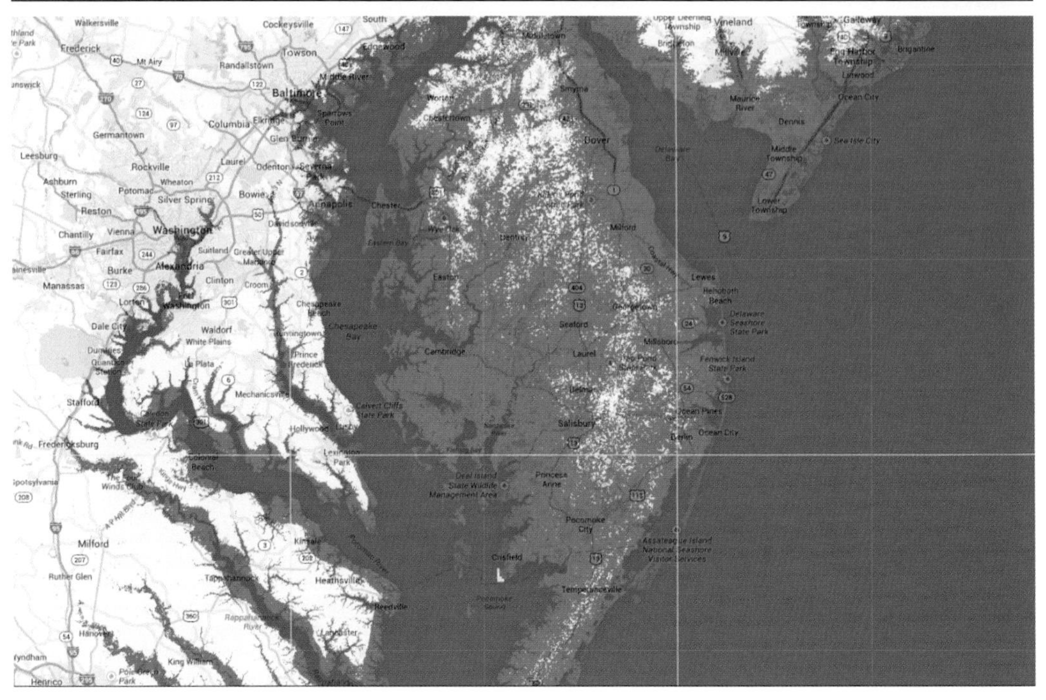

FIGURE B The long-term worst-case sea-level rise from ice sheets could be more than 60 m if all of Greenland and Antarctic ice melts. A 20 m rise, equivalent to loss of all of Greenland's ice, all of the ice in West Antarctica, and some coastal parts of East Antarctica, is shown here. This may approximate the sea level during the Pliocene period (3–5 million years ago), the last time that CO_2 levels are thought to have been 400 ppm. This figure emphasizes the large areas of coastal infrastructure that are potentially at risk if substantial ice sheet loss were to occur. SOURCE: http://geology.com/sea-level-rise/washington.shtml.

for the Pliocene Epoch, 3 to 5 million years ago (see Figure 2.5), when CO_2 is estimated to have had levels similar to today of roughly 400 parts per million, most of Delaware, the first State in the Union, would be under water without very large engineering projects (Figure B). In terms of the expected lifetime of a State, this could also qualify as abrupt.

[a] http://stateofthecoast.noaa.gov/coastal_economy/.
[b] http://www.bls.gov/cew/.
[c] http://www.bls.gov/cew/.
[d] http://www.bls.gov/cew/.
[e] http://www.bls.gov/cew/; http://data.worldbank.org/indicator/NY.GDP.MKTP.CD.
[f] http://factfinder2.census.gov/faces/nav/jsf/pages/index.xhtml.
[g] http://factfinder2.census.gov/faces/nav/jsf/pages/index.xhtml.
[h] http://factfinder2.census.gov/faces/nav/jsf/pages/index.xhtml.
[i] Woods and Poole Economics Inc., 2011; http://coastalsocioeconomics.noaa.gov/.

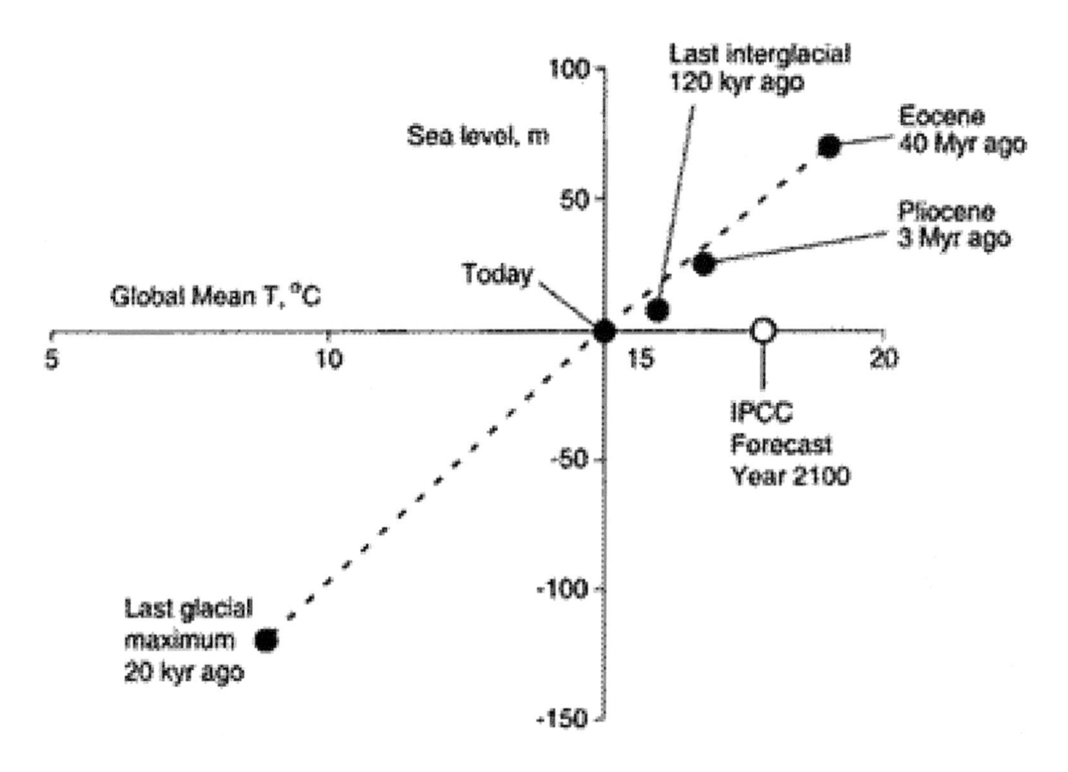

FIGURE 2.5 Co-variation of sea level with global average temperature in the geologic past, compared with the IPCC forecast for sea level rise by the year 2100. Figure from Archer, 2010. The straight line shown may be less accurate than a curve passing through each of the solid dots shown (e.g., Gasson et al., 2012), but sufficient warmth gives large sea-level rise in models and has done so in Earth's history.

tions for this century, although on the high side; Moore et al., 2013) sustained over 33 years would be abrupt for an airport runway that is built to last longer than that (and financed over that time period), but becomes unusable with a half meter of sea-level rise (assuming no adaptive measures are taken).

Rise of the global average sea level over the time periods of most interest to human economies is controlled primarily by the mass or density of ocean water. Local relative sea level may be affected by several additional factors (Box 2.2).

For global average sea level, the main control on water density over these times is ocean temperature, with warming causing thermal expansion by roughly 0.4 m per degree C (Levermann et al., 2013). In response to atmospheric warming, the temperature of the bulk of the ocean will increase primarily through downward transport of water heated at the surface. Because the time for water to move through the deep

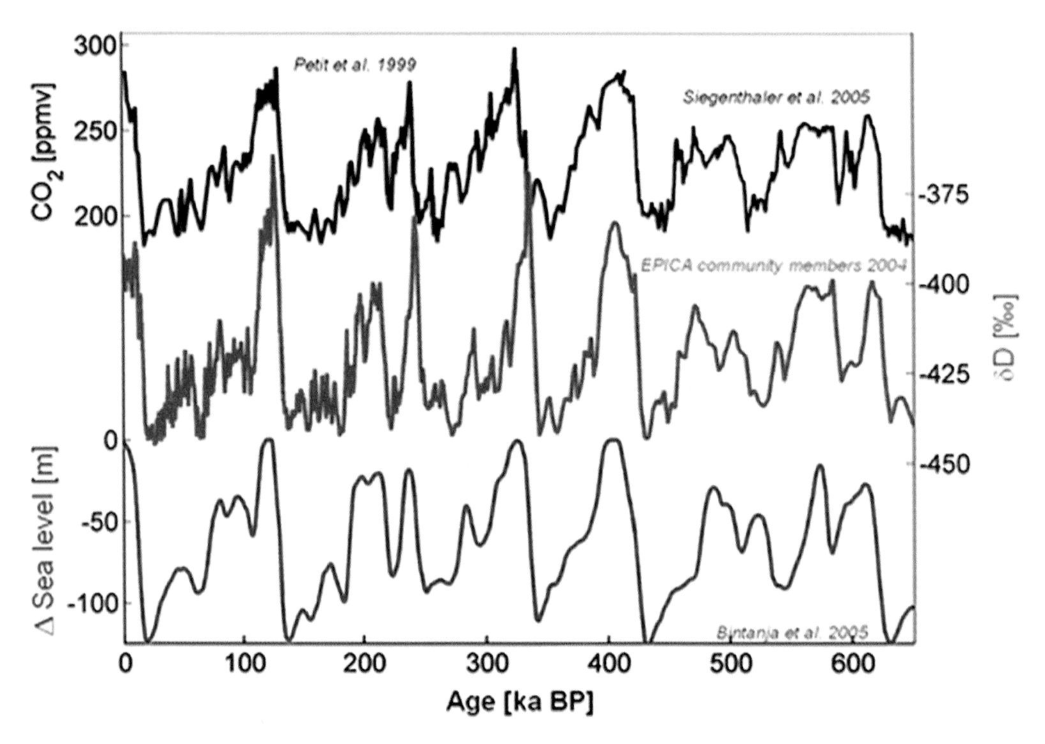

FIGURE 2.6 Sea level and climate over the last 600,000 years. Compilation of Vostok and EPICA Dome C CO_2 concentrations (Petit et al., 1999; Siegenthaler et al., 2005) and δD (deuterium isotope record) as a proxy for local air temperature (Petit et al., 1999; Augustin et al., 2004) and the changes in global sea level relative to the present level (Bintanja et al., 2005). To a first approximation, sea-level changes reflect the volume of ocean water bound in continental ice sheets during the ice ages. CO_2 minima were reached approximately when the sea level was at a minimum, hence, the extent of the continental ice sheets were at a maximum, and the highest CO_2 levels were found during interglacials during the high stands of the sea level. The evolution of the local temperature (as deduced from δD) follows this overall picture and points to a strong coupling of the climate and the carbon cycle. Source: http://www.awi.de/en/research/research_divisions/geosciences/glaciology/ techniques/high_precision_d13c_and_co2_analysis/.

ocean is of the order of 1000 years, thermal expansion is expected to be relatively slow and predictable, although shifts in ocean circulation can influence the details of the warming and sea-level rise. Ocean mass over human time-scales is affected primarily by shifting water between ocean and land. Water may be stored on land in lakes or glaciers, and in spaces in the earth (whether the water is frozen in permafrost, or liquid as groundwater). Extraction of water from the ground for human use may exceed natural recharge, and that water quickly reaches the ocean (Lowe and Gregory, 2006; Headly and Severinghaus, 2007), raising sea level. However, increased storage in

BOX 2.2 LOCAL SEA-LEVEL CHANGE

Local sea-level change can differ notably from the global average for multiple reasons. Changes in land elevation may occur in response to many processes, including mountain-building (tectonic) processes, or flow or bending of rocks caused by ongoing or past changes in loading from ice, water or sediment (isostatic changes). In addition, compaction following removal of groundwater or fossil fuels, or possibly inflation from injection of fluids, may change land elevation (e.g., Bindoff et al., 2007; Sella et al., 2007). These changes can locally accelerate, reduce, or even reverse the global-mean trend, especially if the trend is not too much larger than that observed recently.

In addition, the ocean surface is not level, but exhibits topography caused by winds piling up water along some coast lines and pushing it offshore in other areas, and because different places have water with different local temperature and salinity, and thus density. For rising-CO_2 model experiments tracking warming of the ocean with changing water density and winds, some sites exhibit simulated local sea-level rise that is twice the global-average rise, and other sites exhibit no rise (e.g., Lowe and Gregory, 2006).

Also important is the self-gravitation of ice sheets (e.g., Mitrovica et al., 2001). All masses are gravitationally attracted to other masses. The great bulks of the Greenland and Antarctic ice sheets actually have pulled ocean water toward them, so that their coastal sea levels are notably higher than they would be without that gravitational attraction. If the ice melts, adding water to the ocean, it is useful then to think of a two-step process (see Figure A), although the steps are coupled. First, the world ocean rises rapidly and nearly uniformly, within months or years, as the water from the melting ice spreads around the globe. Second, the ocean near the melting ice sheet drops because the smaller ice-sheet mass has less gravitational attraction for ocean water than before, and thus the water released from the former gravitational attraction of the ice sheet causes additional sea-level rise far from the ice sheet. Very near an ice sheet, the second effect may be larger than the first, with ice-sheet melting causing sea-level fall. Only a tiny fraction of people live near the world's great ice sheets, and for most of the world's coastlines the resulting local rise in sea level is larger than the global average, perhaps approaching 50 percent faster than the global average. Changes in Earth's rotation from the redistribution of mass as the ice melts and ocean responds also contribute slightly to local deviations from the global average.

As discussed in the main text, however, if the large ice sheets were to begin to melt rapidly, the influence of this water being added to the ocean could greatly exceed all of these other effects except for self-gravitation for the vast majority of coastal sites. In such circumstances, almost all coastal sites removed from the immediate vicinity of the melting ice sheet would expect local sea-level rise proportional to the ice-sheet melting.

continued

BOX 2.2 CONTINUED

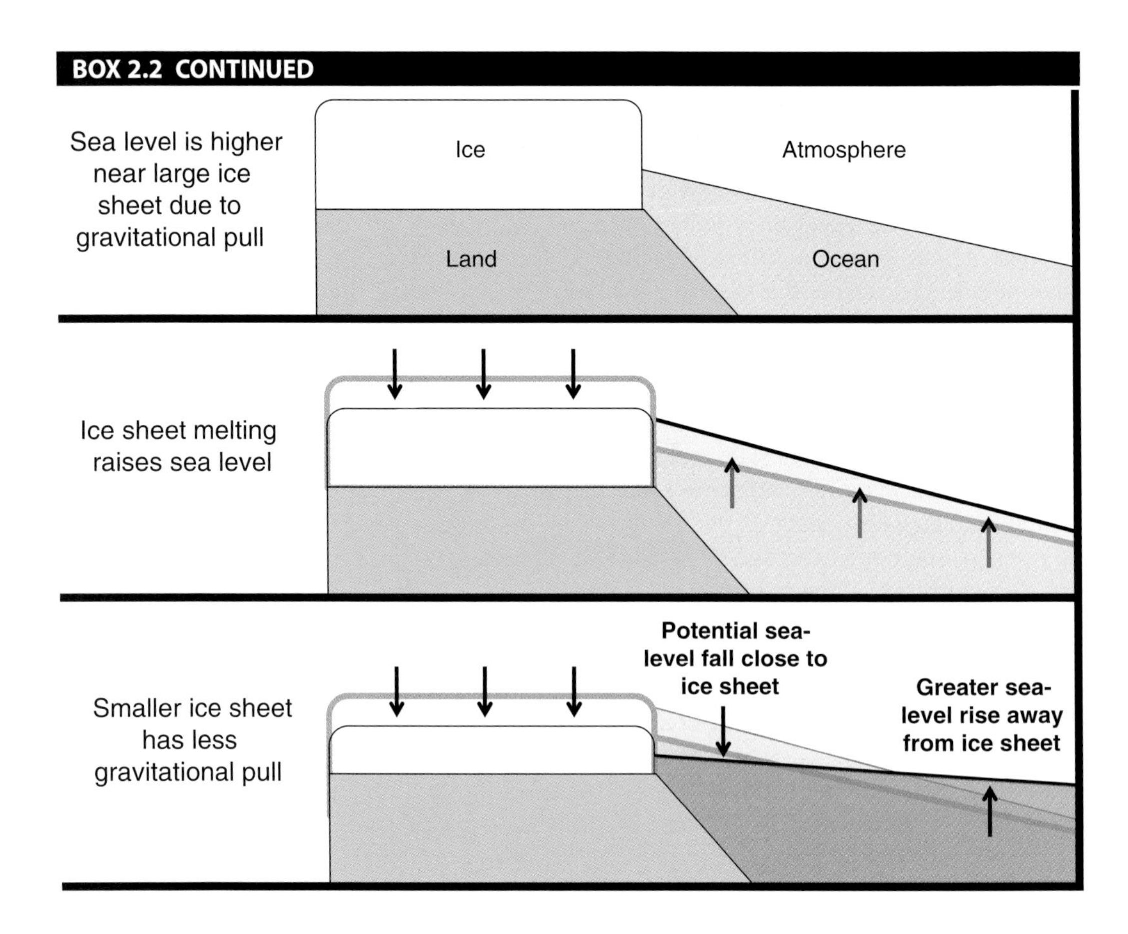

FIGURE A Effect of "self-gravitation" of ice sheets on sea-level rise during ice-sheet melting. The gravitational attraction from the mass in an ice sheet causes sea level to be higher near the ice than the global average. Melting of ice raises the global average sea level, and reduces the gravitational attraction from the ice, which allows the sea level near the ice to fall while sea level far from the ice rises more than the global average.

BOX 2.2 CONTINUED

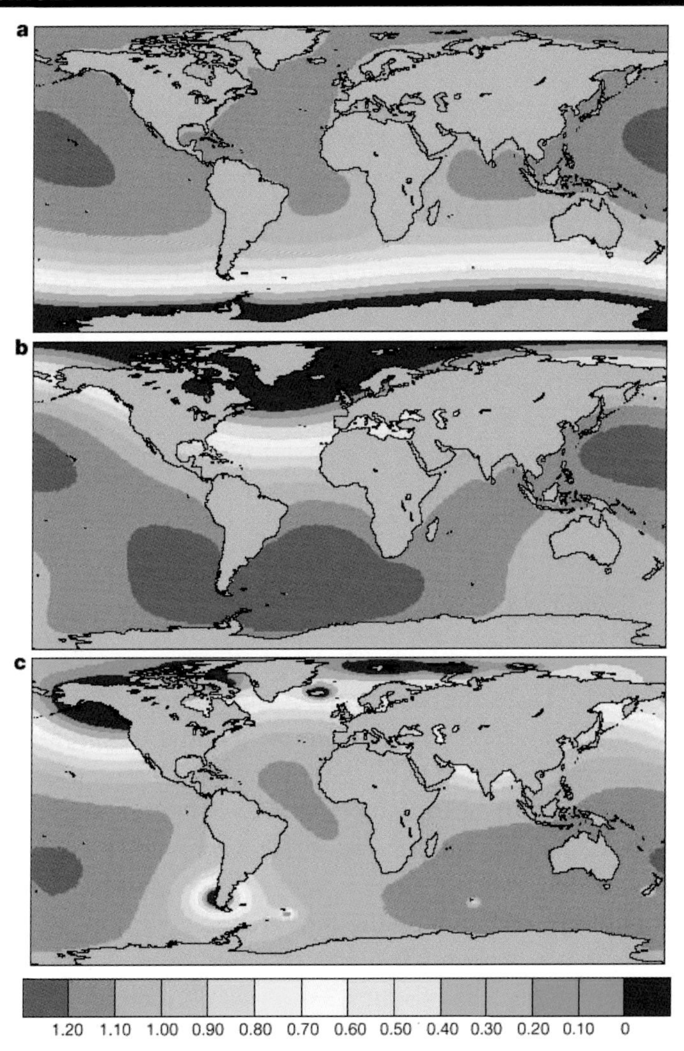

FIGURE B Predicted geometries of sea-level change due to continuing ice mass variations. Values shown are ratios of sea-level rise from enhanced analysis to those from a uniform sea-level rise. These normalized global sea-level variations were computed for the case of present-day ice mass variations in (a) Antarctica and (b) Greenland, as well as (c) melting of the mounting glaciers and ice sheets tabulated by Meier. In (a) and (b) it is assumed that the mass variation is uniform over the two polar regions. The results are normalized by the equivalent eustatic sea-level change for each mass flux event (see original reference). Departures from a contour value of 1.0 reflect departures from the assumption that the sea-level distribution accompanying these mass flux events is uniform. Predictions are based on a new sea-level theory solved using a pseudo-spectral algorithm with truncation at spherical harmonic degree and order 512. This truncation corresponds to a spatial resolution of 40 km. Source: Mitrovica et al., 2001.

new human-made impoundments or in groundwater fed by such impoundments will lower sea level. These partially offsetting effects lead to the expectation that direct human shifts in water storage on land will not have large effects on sea level in comparison to the effects of ocean warming and mountain-glacier and ice-sheet melting (Wada et al., 2012), although notable uncertainties remain in regards to future groundwater use and reservoir construction, and these effects vary considerably depending on the specific location (NRC, 2012e).

Most mountain glaciers worldwide are losing mass, contributing to sea-level rise. However, the amount of water stored in this ice is estimated to be less than 0.5 m of sea-level equivalent (Lemke et al., 2007), so the contribution to sea-level rise cannot be especially large before the reservoir is depleted. On the other hand, the reservoir in the polar ice sheets is sufficient to raise global sea level by more than 60 m (Lemke et al., 2007). Thus any large and rapid global sea-level rise, if it were to occur, would almost surely be sourced from the ice sheets.

Potential Abrupt Changes to Polar Ice Sheets

Ice-sheet volume is controlled by the balance between mass input and mass loss; mass input is almost entirely due to snowfall, and mass loss is from iceberg calving supplied by flow of the ice sheet, or runoff of melt water. As summarized in, for example, Meehl et al. (2007b), warming is expected to increase snowfall in the colder parts of Greenland and in Antarctica, and to increase melting in the warmer parts of Greenland. Beyond some threshold of a few degrees C warming, Greenland's ice sheet will be almost completely removed. However, the timescale for this is expected to be many centuries to millennia, depending somewhat on the model used and more strongly on the emissions pathway (e.g., Meehl et al., 2007b). This still could result in a relatively rapid rate of sea-level rise. Greenland ice has about 7.3 m equivalent of sea-level rise (Lemke et al., 2007), which, if melted over 1000 years (a representative rather than limiting case), yields an annual rise rate of 7 mm/yr just from Greenland, slightly more than twice as fast as the recent rate of rise from all sources including melting of Greenland's ice. Slower melting would obviously yield lower average rates, but the potential for rapid rise still exists. Surface melting removing the Antarctic ice sheet would require much more warming than in Greenland.

The loss of land ice by melting would be reversible if sufficient cooling were applied sufficiently rapidly (Ridley et al., 2010). For example, if the temperature increased across the threshold needed to remove the modern Greenland ice sheet, almost immediate cooling even slightly below the threshold would allow ice-sheet persistence.

Ice-sheet shrinkage with continuing melting lowers the surface into warmer parts of the atmosphere and warms the surroundings by replacing reflective snow and ice with dark rock, thus requiring greater cooling to allow regrowth.

Mass loss by flow of ice into the ocean is less well understood, and it is arguably the frontier of glaciological science where the most could be gained in terms of understanding the threat to humans of rapid sea-level rise. Increased ice-sheet flow can raise sea level by shifting non-floating ice into icebergs or into floating-but-still-attached ice shelves, which can melt both from beneath and on the surface. Rapid sea-level rise from these processes is limited to those regions where the bed of the ice sheet is well below sea level and thus capable of feeding ice shelves or directly calving icebergs rapidly, but this still represents notable potential contributions to sea-level rise, including the deep fjords in Greenland (roughly 0.5 m; Bindschadler et al., 2013), parts of the East Antarctic ice sheet (perhaps as much as 20 m; Fretwell et al., 2013), and especially parts of the West Antarctic ice sheet (just over 3 m; Bamber et al., 2009).

In understanding the behavior of ice sheets, attention is particularly focused on the boundary between the floating ice and grounded ice, which is usually called the grounding line, although in detail it is a zone with interesting but imperfectly understood properties (e.g., Schoof, 2007; Joughin et al., 2012a; Walker et al., 2013); see Figure 2.7. Large changes in ice mass are generally tightly coupled to grounding-line migration. If the ice-sheet bed deepens toward the center of the ice sheet, an instability exists, such that in the absence of additional stabilizers, the grounding line will advance with ice-sheet growth, or retreat with ice-sheet shrinkage, to a position where the bed rises towards the ice-sheet center. This instability can be overcome by a local reversal of the bed or narrowing of a fjord, and especially by friction between ice shelves and fjord walls or local highs in the sea floor.

The important role of ice shelves in stabilizing marine ice sheets introduces the potential for large and rapid ice-sheet shrinkage. The warmest upper surfaces of ice sheets are generally on ice shelves, because they are the lowest-elevation parts, and extend away from the cold central regions towards generally warmer oceans. Where meltwater forms on the ice-shelf surface, it can wedge open crevasses and cause ice-shelf disintegration, much like a line of balanced dominoes falling over, which has been observed to occur within weeks in the rapidly warming Antarctic Peninsula region (e.g., MacAyeal et al., 2003). Ice shelves are in contact with ocean water, and any warming of the water or increase in circulation of warm water under the shelves contributes to faster melting. Thinning or loss of ice shelves reduces friction, allows faster flow of the non-floating ice feeding the shelves, and thus contributes to sea-level rise. Furthermore, the recent behavior of ice shelves in Greenland (Nettles and Ekstrom, 2010),

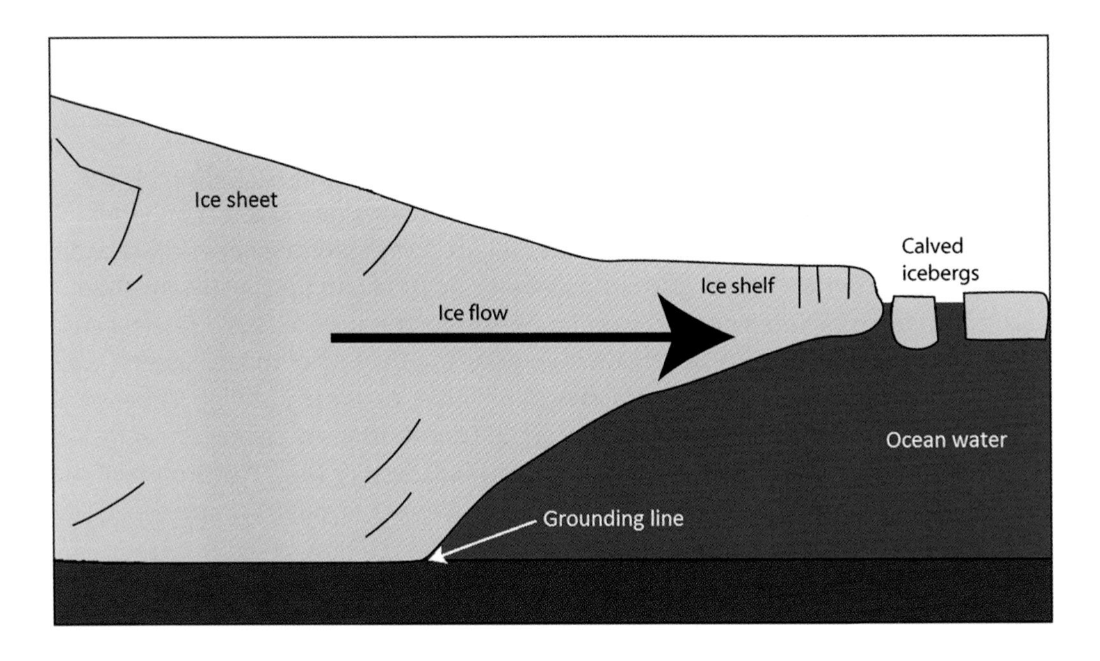

FIGURE 2.7 Schematic showing the grounding line for an ice sheet.
SOURCE: Adapted from www.AntarcticGlaciers.org by Bethan Davies. Used with permission.

the sedimentary record (Jakobsson et al., 2011), and our physical understanding (Alley et al., 2008) suggest that beyond some threshold ice-shelf reduction leads to complete loss as the ice shelf calves away, potentially in less than or much less than one year.

Model results, and the sedimentary record of deglaciated marine regions in both hemispheres, show that grounding lines tend to stabilize on local seafloor highs or fjord narrowings, and then to increase that stability through sedimentation (Anandakrishnan et al., 2007; Alley et al., 2007; Dowdeswell et al., 2008). In this stabilized position, the grounding line is little-affected by sufficiently small environmental forcings (e.g., Horgan and Anandakrishnan, 2006). However, for forcing beyond some threshold, the grounding line migrates rapidly to a new position of stability, which may be far away and involve an important contribution to sea-level change.

The possible rates of this migration are poorly quantified. The changes occurring now in drainages including Jakobshavn Glacier in Greenland (e.g., Joughin et al., 2012b) may be analogous to the events during retreat following the last ice age, but there at least are suggestions that past changes were sometimes faster (Jakobsson et al., 2011). The rate of discharge into deep water across a grounding line in general increases with the thickness of the ice and the width of the fjord or channel through which the ice discharges. A retreat of Thwaites Glacier in West Antarctica could give a much wider and deeper calving front than any observed today, so the "speed limits" suggested by Pfeffer et al. (2008) may not apply (Parizek et al., 2013).

Because the full suite of physical processes at the grounding line (e.g., Walker et al., 2013) in general is not represented in modern models, the possibility exists that rates produced by extant models under strong simulated forcing may be greatly in error (Nowicki et al., 2013). Deglaciation of the marine portions of West Antarctica would raise sea level by ~3.3 m (Bamber et al., 2009), with additional marine ice in East Antarctica and Greenland, as noted above. Improved understanding of the retreat rates of WAIS and other marine based ice drainage zones is necessary to narrow the currently broad uncertainties and better quantify the potential worst-case scenarios.

Much process-based research coupling field work, remote sensing, and modeling is required to advance assessment of the likelihood of a threshold-crossing leading to abrupt sea-level rise from the ice sheets, as well as to improve projections of more-gradual sea-level rise that could lead to threshold-crossing events in other systems. Great progress has been made recently in assessing the current rate of mass loss from the ice sheets (Shepherd et al., 2012), as well as monitoring the changing snowfall, surface melting, and temperature contributing to the changes. But maintenance and expansion of this effort are threatened, especially by possible loss of satellite observing capabilities (NRC, 2012d). The strong coupling of ice sheets and surrounding oceans (Joughin et al., 2012a) was not fully anticipated in early modeling efforts, and is not now fully represented in comprehensive Earth-system models. Key environmental information includes air temperatures and ocean temperatures in the upper kilometer of the ocean, sea ice, and related oceanic properties. Air temperature is relatively well monitored, although observations in remote polar areas are not dense, and that lack of data density can be problematic. More fixed monitoring sites as well as UAV-based observations are needed in the remote areas of both poles. Ocean temperatures are not well monitored, particularly in polar regions and particularly near the grounding lines and along the ice-ocean interface for marine-based ice. A concerted effort is needed to collect better data for constraining ocean conditions.

The second part of monitoring is to continually catalog those sensitive parts of coastal systems, human and natural, that are vulnerable to the impacts of a slowly or rapidly rising sea level and may exhibit abrupt behavior in response to a rising sea level, as well as the local changes in sea level. New surveys should be a routine part of environmental monitoring. Given the slow speed of sea-level rise, such new surveys need not be annual, but regular surveys are needed at a frequency that takes into account local building rates as well as local rates of sea-level change.

Summary and the Way Forward

Sea level is rising, primarily in response to a warming planet, through thermal expansion of the oceans, and also via the loss of land ice as ocean and air temperatures increase, melting ice and speeding the flow of non-floating ice to form floating icebergs. Thermal expansion is expected to continue, and to be a slow but steady process. Loss of land ice by direct melting is also expected to be slow and steady. Both of these processes are likely to contribute to abrupt change via a slowly rising sea level forcing other systems to cross thresholds. Examples of such systems include human-built infrastructure at or near the coasts, natural ecosystems, freshwater in the ground, and natural dams or topographic choke points. Storm surges are exacerbated by rising sea level, amplifying the extent of the problem. Moving forward, tracking sea-level rise will require maintenance and expansion of the monitoring of sea level (tide gauges and satellite data), ocean temperatures at depth, and local coastal motions.

Sea level itself may also increase rapidly enough to be termed abrupt (O'Leary et al., 2013). The loss of land ice, particularly from marine-based ice sheets such as the West Antarctic Ice Sheet—possibly in response to gradual ocean warming—could trigger sea-level rise rates that are much higher than ongoing. Paleoclimatic rates at least 10 times larger than recent rates have been documented, and similar or possibly higher rates cannot be excluded in the future. This time scale is also roughly that of human-built infrastructure such as roads, water treatment plants, tunnels, homes, etc. Deep uncertainty persists about the likelihood of a rapid ice-sheet "collapse" contributing to a major acceleration of sea-level rise; for the coming century, the probability of such an event is generally considered to be low but not zero (e.g., Bamber and Aspinall, 2013). To reduce this uncertainty moving forward, extensive effort is required, including the maintenance and expansion of satellite measurements, aerogeophysical monitoring, surface monitoring, process studies, and modeling.

Because air carries much less heat than an equivalent volume of water, physical understanding indicates that the most rapid melting of ice leading to abrupt sea-level rise is restricted to ice sheets flowing rapidly into deeper water capable of melting ice rapidly and carrying away large volumes of icebergs. In Greenland, such deep water in contact with ice is restricted to narrow bedrock troughs where friction between ice and fjord walls limits discharge. Thus, the Greenland ice sheet is not expected to destabilize rapidly within this century. However, a large part of the West Antarctic

Ice Sheet (WAIS), representing 3-4 m of potential sea-level rise, is capable of flowing rapidly into deep ocean basins. Because the full suite of physical processes occurring where ice meets ocean is not included in comprehensive ice-sheet models, it remains possible that future rates of sea-level rise from the WAIS are underestimated, perhaps substantially. Improved understanding of key physical processes and inclusion of them in models, together with improved projections of changes in the surrounding ocean, are required to notably reduce uncertainties and to better quantify worst-case scenarios. Because large uncertainties remain, the committee judges an abrupt change in the WAIS within this century to be plausible, with an unknown although probably low probability.

Changes in Ocean Chemistry and Associated Effects on Marine Ecosystems

Increasing carbon dioxide concentrations in the atmosphere, changing windstress, currents and mixing, and increasing temperatures are changing the chemistry of the world's oceans. These changes are resulting in a decrease in pH, carbonate ion concentrations, and dissolved oxygen in the ocean.

Ocean Acidification

Carbon dioxide combines with water to form carbonic acid, which then dissociates to form bicarbonate ions and hydrogen ions (H^+), so that increasing concentrations of CO_2 in the atmosphere have been decreasing the pH (acidifying) of the surface ocean (NRC, 2010c). Since the preindustrial period, pH has declined by approximately 0.1 pH unit corresponding to a 30 percent increase in acidity. By 2100, the acidity is projected to increase by 100-150 percent compared to preindustrial values. Geologic records indicate that when the increase in atmospheric CO_2 is gradual, oceanic pH and carbonate levels remain relatively stable due to processes that occur in equilibrium, such as dissolution of $CaCO_3$ shells, weathering of terrestrial rock, and tectonic processes. However, the current rate of increase of atmospheric CO_2 exceeds the rate at which natural processes can buffer these pH changes.

Although ocean acidification is not an abrupt climate change, i.e., the change occurs at the same rate as the forcing, the impacts of ocean acidification on ocean biology have the potential to cause rapid (over multiple decades) changes in ecosystems and to be irreversible when contributing to extinction events. Specifically, the increase in CO_2 and HCO_3^- availability might increase photosynthetic rates in some photosynthetic marine organisms, and the decrease in CO_3^{2-} availability for calcification makes it increasingly difficult for calcifying organisms (such as some phytoplankton, corals, and bivalves) to build their calcareous shells and effects pH sensitive physiological processes (NRC, 2010c, 2013). As such, ocean acidification could represent an abrupt climate impact when thresholds are crossed below which organisms lose the ability to create their shells by calcification, or pH changes affect survival rates (see the Extinctions section below for more discussion of these issues).

Scientists are particularly concerned with the ability of reef-building corals to persist in the face of ocean acidification combined with the other stressors such as temperature increase, sea level rise, and changes in storm intensity all also associated with climate change. In addition, eutrophication and physical injuries inflicted from SCUBA diving and fishing further impact the health of coral ecosystems. Cascading events could irreversibly alter reef ecosystems on short time scales. However, the understanding of the effects on marine ecosystems is too limited to be able to draw any conclusions about the magnitude and rate of changes to come.

In addition, combined with the decline in oxygen availability, ocean acidification has the potential to impair aerobic respiration (see further discussion below). Changes in near-coastal circulation or biochemistry seem to be altering surface ocean pH more quickly than can be explained by an equilibrium response to the rising atmospheric CO_2 concentration (Wootton and Pfister, 2012). This topic requires further research (see discussion below in this section).

Oxygen Content in the Global Ocean

The oxygen content in the surface ocean is projected to decline with warming because of the decrease in solubility of gases with increasing temperature, and changes in ventilation and biological consumption. A significant decrease in oxygen in the upper ocean between the 1970s and 1990s has already been observed at a global scale (Helm et al., 2011). Only approximately 15 percent of that decline can be attributed to a warmer mixed-layer, with the remainder being "consistent with an overall decrease in the exchange between surface waters and the ocean interior" (Helm et al., 2011). With a general weakening of ventilation rates as a result of climate change (Bryan et al., 2006), oxygen content of the global ocean is likely to further decrease (ventilation to the surface allows new input of oxygen from the atmosphere).

Of more immediate concern is the expansion of Oxygen Minimum Zones (OMZs). Photosynthesis in the sunlit upper ocean produces O_2, which escapes to the atmosphere; it also produces particles of organic carbon that sink into deeper waters before they decompose and consume O_2. The net result is a subsurface oxygen minimum typically found from 200–1000 meters of water depth, called an Oxygen Minimum Zone.

Warming ocean temperatures lead to lower oxygen solubility. A warming surface ocean is also likely to increase the density stratification of the water column (i.e., Steinacher et al., 2010), altering the circulation and potentially increasing the isolation of waters in an OMZ from contact with the atmosphere, hence increasing the intensity of the OMZ. Thus, oxygen concentrations in OMZs fall to very low levels due to the consumption of organic matter (and associated respiration of oxygen) and weak replenishment of oxygen by ocean mixing and circulation. Furthermore, a hypothetical warming of 1°C would decrease the oxygen solubility by 5 µM (a few percent of the saturation value). This would result in the expansion of the hypoxic[2] zone by 10 percent, and a tripling of the extent of the suboxic zone (Deutsch et al., 2011). With a 2°C warming, the solubility would decrease by 14 µM resulting in a large expansion of areas depleted of dissolved oxygen and turning large areas of the ocean into places where aerobic life disappears. In the tropical Atlantic, Pacific, and Indian Ocean, a decline in oxygen content in the subsurface waters has been confirmed with observations (Stramma et al., 2010).

The expansion and intensification of existing OMZs and the increase in CO_2 are likely to pose a threat to aerobic marine life (Brewer and Peltzer, 2009). The amount of dissolved oxygen that marine life requires depends on how oxygen can diffuse across tissue boundaries, which is a function of oxygen content, temperature, and pressure. Some researchers have suggested that a respiration index (RI) be defined—based on the ratio of the partial pressures of oxygen and CO_2—as a better metric for estimating the physiological limits of deep sea animals (Brewer and Peltzer, 2009). The use of this particular respiratory index has been disputed (Seibel and Childress, 2013), but it would be useful to develop a metric that could allow for a better assessment of the global extent of water masses where aerobic organisms could not survive. It could also contribute to improving early detection of thresholds for mass mortalities of aerobic organisms, which is of particular importance considering the economic value the fishing industry.

Limits to aerobic life in the sea are often defined as ~5 µM, below which it is inefficient for aerobic microbes to consume dissolved oxygen (Brewer and Peltzer, 2009). While some species adapted to lower-oxygen conditions, paleo records have shown the extinctions of many benthic species during past periods of hypoxia. These periods have

[2] Hypoxia is the environmental condition when dissolved water column oxygen (DO) drops below concentrations that are considered the minimal requirement for animal life. Suboxia is even further depletion of oxygen and anoxia is the condition of no oxygen at all.

coincided with both a rise in temperature and sea level. Records also indicate long recovery times for ecosystems affected by hypoxic events (Danise et al., 2013).

In addition, when the oxygen in seawater is depleted, bacterial respiration of organic matter turns to alternate electron-acceptors with which to oxidize organic matter, such as dissolved nitrate (NO_3^-). A by-product of this "denitrification" reaction is the release of N_2O, a powerful greenhouse gas with an atmospheric lifetime of about 150 years. Low-oxygen environments, in the water column and in the sediments, are the main removal mechanism for nitrate from the global ocean. An intensification of oxygen depletion in the ocean therefore also has the potential to alter the global ocean inventory of nitrate, affecting photosynthesis in the ocean. However, the lifetime of nitrate in the global ocean is thousands of years, so any change in the global nitrate inventory would also take place on this long time scale.

Likelihood of Abrupt Changes

Changes in global ocean oxygen concentrations have the potential to be abrupt because of the threshold to anoxic conditions, under which the region becomes uninhabitable for aerobic organisms including fish and benthic organisms. Once this tipping point is reached in an area, anaerobic processes would be expected to dominate resulting in a likely increase in the production of the greenhouse gas N_2O. Some regions like the Bay of Bengal already have low oxygen concentrations today (Delaygue et al., 2001), but not quite low enough for denitrification to occur. Modest increases in the export of organic matter, or decreases in ventilation by the circulation, could decrease oxygen below the critical threshold for fixed nitrogen loss.

OMZs have also been intensified in many areas of the world's coastal oceans by runoff of plant fertilizers from agriculture and incomplete wastewater treatment. These 'dead zones' have spread significantly since the middle of the last century and pose a threat to coastal marine ecosystems (Diaz and Rosenberg, 2008). This expansion of OMZs is due to nutrient runoff makes the ocean more vulnerable to decreasing solubility of O_2 in a warmer ocean. Indeed, as warming of the ocean intensifies, the decrease in oxygen availability might become non-linear; particularly, as indicated by the expansion of the size of the oxygen minimum zone (Deutsch et al., 2011). The effect of temperature on oxygen solubility is well understood. However, it remains a major scientific challenge to model and project the changes of the magnitude and intensity of subsurface oxygen depletion because it depends on changes in ocean circulation, rates of de-nitrification, and nutrient runoff from land, and because global data coverage for chemical and biological parameters remains poor.

Summary and the Way Forward

In order to understand and possibly anticipate changes to the chemistry of the world's oceans, the oxygen content, pH, and temperature of subsurface waters need to be monitored at the global, synoptic scale. The majority of the available oxygen data stem from the coastal oceans or from the World Ocean Circulation Experiment (WOCE) that took shipboard measurements across large portions of the ocean. However, the data remain too sparse in time and space (Stramma et al., 2010) to be able to detect long-term trends with confidence.

As oxygen sensors have become more sophisticated and accurate, they can be deployed more widely on buoys and floats. The current monitoring effort would ideally be expanded to equip Argo floats with oxygen sensors to achieve more global coverage in oxygen data. In order to better understand the effects of ocean chemistry on marine ecosystems, oxygen monitoring needs to be supplemented with biological observations at some select sites.

OMZs are not well represented in global climate models due to limited understanding of the physical and biological processes that affect them. In particular, the processes that lead deep water to be exchanged with the surface water remain poorly understood; for example, how rapidly a given parcel of ocean water is ventilated needs to be better resolved. Understanding such processes would enable models to be improved. Thus, physical processes such as vertical and isopycnal mixing that drive large scale circulations need to be better understood to improve the predictive capability and accuracy of the models.

Furthermore, research would benefit from new and standardized methods. For example, oxygen data need to be accompanied by contemporaneous pressure and temperature data as these variables combined give a better indication of how readily oxygen can diffuse across tissue boundaries. In addition, scientists could benefit by using common definitions for hypoxia, suboxia, and anoxia (Hofmann et al., 2011). While much research on the effects of shallow coastal dead zones has been published, little is understood on how this expansion will affect open ocean ecosystems.

Lastly, biological processes need to be better understood, including the microbial processes in OMZs, as well as how much larger organisms are affected and can adapt to the changes in OMZs. Resolving these questions would require a major effort given that OMZs represent relatively remote and under-sampled areas of the ocean.

ABRUPT CHANGES IN THE ATMOSPHERE

Atmospheric Circulation

The climate system exhibits variability on a range of spatial and temporal scales. On large (i.e., continental) scales, variability in the climate system tends to be organized into distinct spatial patterns of atmospheric and oceanic variability that are largely fixed in space but fluctuate in time. Such patterns are thought to owe their existence to internal feedbacks within the climate system.

Prominent patterns of large-scale climate variability include:

- the El-Nino/Southern Oscillation (ENSO),
- the Madden-Julian Oscillation (MJO),
- the stratospheric Quasi-Biennial Oscillation,
- the Pacific-North American pattern, and
- the Northern and Southern annular modes (the Northern annular mode is also known as the North Atlantic Oscillation).

All have a pronounced signature in atmospheric variability, and all owe their existence to internal climate dynamics. For example, ENSO is characterized by episodic warming and cooling of the eastern tropical Pacific, owes its existence to feedbacks between the tropical ocean and atmosphere, and fluctuates on timescales of ~2-7 years. The annular modes are characterized by north-south vacillations in the jetstream at middle latitudes, owe their existence to internal atmospheric dynamics, and fluctuate on timescales spanning weeks to decades. (In the discussion that follows, the middle latitude jetstreams are the eastward flowing air currents centered in middle latitudes near 6 to 12 km. The jetstreams are frequently collocated with wintertime storms.)

Abrupt climate change due to variations in the atmospheric circulation and its attendant patterns of climate variability can arise through two principal mechanisms: (1) through abrupt changes in the time-dependent behavior of the circulation; or (2) through slowly evolving changes in the circulation that project onto large horizontal gradients in surface weather. For example, a relatively slow shift in the distribution of precipitation could give rise to relatively rapid changes in precipitation patterns in regions that lie at the interface of dry and rainy regions (see Figure 2.8), potentially altering a location's local climate with possible ramifications to water supplies and/or agriculture for example. The text below discusses the evidence for: (1) abrupt changes in the circulation and (2) steady changes in the circulation that may, in turn, trigger relatively abrupt changes in climate in regions of large spatial gradients in surface weather.

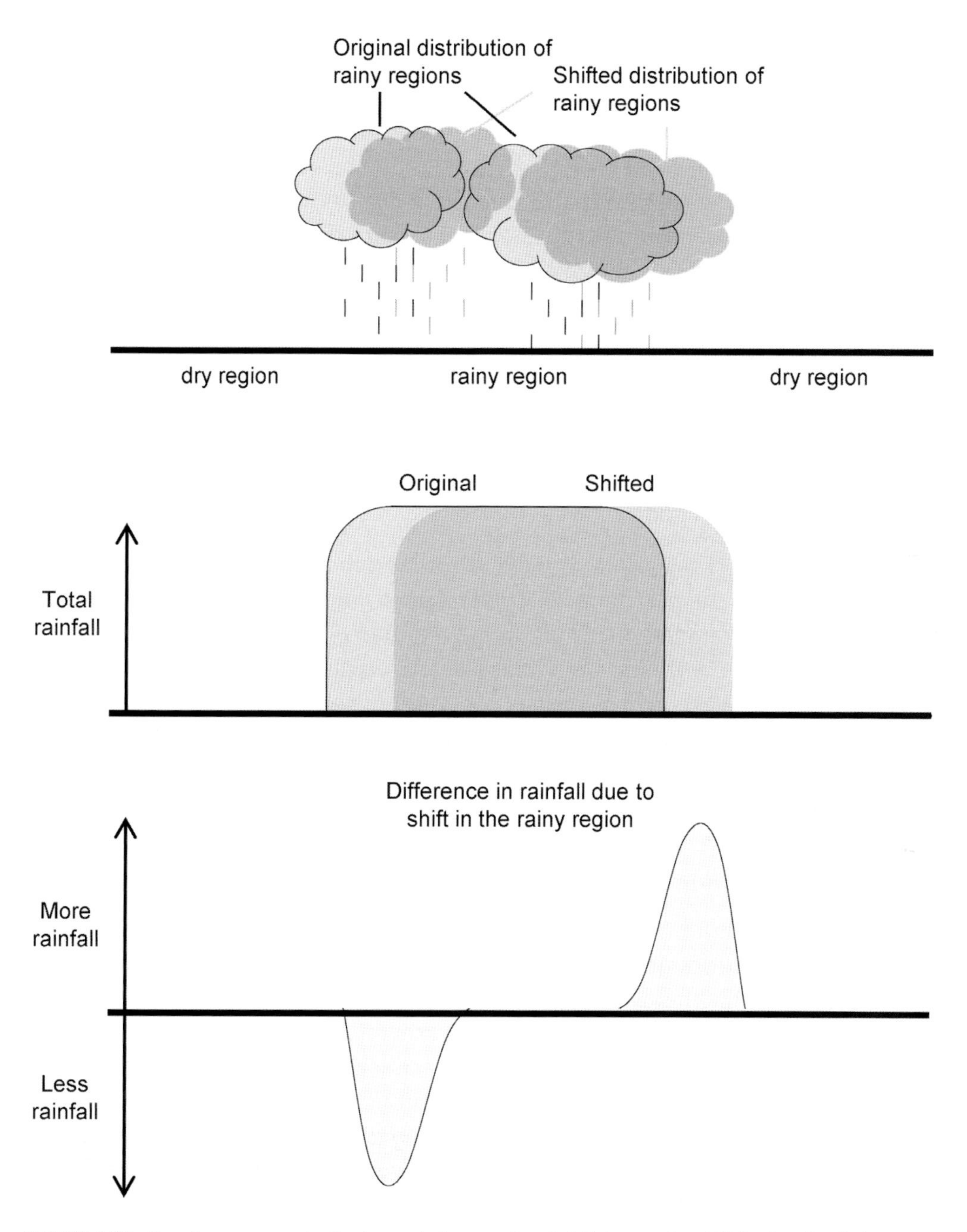

FIGURE 2.8 In the above example, the largest changes in rainfall due to the shift in the circulation are found on the flanks of the original precipitation regions. A slowly evolving change in the circulation may thus lead to seemingly abrupt changes in precipitation in regions where the existing spatial gradients in rainfall are largest.

Abrupt Changes in the Time-Dependent Behavior of the Atmospheric Circulation

Given the definition of abrupt change in this report (see Box 1.2), there is *little* evidence that the atmospheric circulation and its attendant large-scale patterns of variability have exhibited abrupt change, at least in the observations. The atmospheric circulation exhibits marked natural variability across a range of timescales, and this variability can readily mask the effects of climate change (e.g., Deser et al., 2012a, 2012b). As noted above, patterns of large-scale variability in the extratropical atmospheric wind field exhibit variations on timescales from weeks to decades (Hartmann and Lo, 1998; Feldstein, 2000). The time series of large-scale tropical climate variability—such as the MJO and ENSO—exhibit more quasi-periodic behavior (e.g., Rasmusson and Carpenter, 1982; Zhang, 2005). In both the tropics and extratropics, it is difficult to discern significant long-term trends in the patterns of climate variability from natural variability, never mind abrupt (threshold) changes in the atmospheric circulation.

The most widely studied and arguably most robust "regime shift" in the modern historical record (i.e., over the past century) is the relatively rapid change in North Pacific sea-surface temperatures circa 1976, referred to alternatively as ENSO-like decadal variability (Zhang et al., 1997) and the Pacific Decadal Oscillation (PDO; Mantua et al., 1997). Numerous physical mechanisms have been proposed to explain the apparent step-like jump in the extratropical atmosphere/ocean system. But it is unclear whether or not the 1976 regime shift in North Pacific climate reflects an abrupt change in the extratropical atmosphere-ocean system or simply the random superposition of different climate signals, e.g., similar regime-shifts are reproducible in simple stochastic models forced by atmospheric noise and ENSO (Newman et al., 2003).

One recent modeling study indicates that the atmospheric circulation may respond abruptly to future anthropogenic climate change, at least in a simple climate model. Wang et al. (2012b) force the dynamical core of an atmospheric general circulation model with warming in the tropical troposphere that mimics the effects of climate change there. Warmings up to ~5°C lead to steady changes in the atmospheric circulation consistent with those found in full IPCC-class simulations. When the warming is increased beyond 5°C (which is predicted to occur by the end of the 21st century in the IPCC A1B scenario [IPCC, 2007c]), the atmospheric circulation exhibits large and abrupt changes, including a sudden poleward jump in the middle latitude jetstream of roughly 10 degrees latitude. As of this writing, the result is derived from a full primitive equation model, albeit one with very idealized physics. The regime-like behavior found by Wang et al. (2012b) has not been reproduced in a full physics, IPCC-class model simulation.

Steady Changes in the Time-Dependent Behavior of the Atmospheric Circulation

Relatively abrupt changes in the climate of a particular location may be driven not by abrupt changes in the atmospheric circulation, but rather by otherwise slowly evolving changes in the circulation in regions of large horizontal gradients in surface weather. Steady changes in the atmospheric circulation (i.e., changes that scale linearly with the forcing) have been documented in both climate models and observations. The most robust evidence for steady changes in the large-scale atmospheric circulation include:

1. Observational and numerical evidence of a poleward shift in the Southern Hemisphere middle latitude jetstream (a positive trend in the Southern Annular mode) in response to Antarctic ozone depletion (Gillett and Thompson, 2003; Arblaster and Meehl, 2006; Son et al., 2010; Polvani et al., 2011; McLandress et al., 2011; Thompson et al., 2011). The signature of the ozone hole in surface climate is most pronounced during the summer season (Thompson et al., 2011). Opposite signed trends in the Southern Hemisphere middle latitude jetstream are expected in response to the recovery of the Antarctic ozone hole (Son et al., 2010; Arblaster et al., 2011; Polvani et al., 2011). The circulation response to ozone recovery is expected to oppose the response to future increases in greenhouse gases (see 2 below).
2. Numerical evidence of a poleward shift in the Southern Hemisphere and North Atlantic middle latitude jetstreams in response to increasing greenhouse gases (e.g., Fyfe et al., 1999; Kushner et al., 2001; Cai et al., 2003; Yin, 2005; Miller et al., 2006; Meehl et al., 2007b; Barnes and Polvani, 2013). The changes in the flow project strongly onto the Southern annular mode and North Atlantic Oscillation (NAO), respectively. The poleward shift of the Southern Hemisphere middle latitude jetstream in response to increasing carbon dioxide is one of the most robust circulation responses found in climate change experiments, and is predicted to occur during all seasons (IPCC, 2007c). The predicted changes in the Northern Hemisphere circulation are generally much less robust. This is particularly true for the North Pacific (Barnes and Polvani, 2013). Trends in the Northern Hemisphere atmospheric circulation generally do not occur in numerical models until the latter half of the 21st century. The evidence for changes in the circulation in response to increasing greenhouse gases derives primarily from numerical climate model experiments. Observed trends in the middle latitude jetstreams and annular modes are not robust across all months (IPCC, 2007c).
3. Observational evidence and evidence from numerical models for changes in the northern and southern boundaries of the tropics (Fu et al., 2006; Previdi and

Liepert, 2007; Seidel et al., 2008; Lu et al., 2009; Lu et al., 2007; Allen et al., 2012). The observational evidence is on the margins of statistical significance (Davis and Rosenlof, 2012).

4. Numerical evidence for an acceleration of the Brewer-Dobson circulation in response to increasing greenhouse gases (e.g., Butchart et al., 2010; Butchart et al., 2006; Garcia and Randel, 2008; McLandress and Shepherd, 2009; Shepherd and McLandress, 2011; Garny et al., 2011). The Brewer-Dobson circulation is a large-scale circulation cell with rising motion in the tropical stratosphere, poleward motion in the middle latitude stratosphere, and sinking motion in the polar stratosphere. It is the primary pathway for global-scale equator-pole transport in the stratosphere and has implications for ozone concentrations.

Long-term trends in ENSO are uncertain. Lau et al. (2008), Meehl and Teng (2007), and Power and Smith (2007) all argue that future climate change could lead to changes in the amplitude and/or structure of the meteorological anomalies associated with ENSO. But there is little consensus on the response of ENSO itself to climate change and no clear century-scale changes in observed character of ENSO (e.g., Collins et al., 2010; IPCC, 2013; Lenton et al., 2008; Vecchi and Wittenberg, 2010). Observations indicate changes in equatorial Pacific surface pressure (e.g., Bunge and Clarke, 2009; DiNezio et al., 2013; Karnauskas et al., 2009; L'Heureux et al., 2013; Power and Smith, 2007; Vecchi et al., 2006). But the observed trends vary notably depending on the time period being explored, and it is unclear to what extent they reflect internal variability or anthropogenic forcing (e.g., IPCC, 2013; Seager and Naik, 2012).

It is unclear to what extent the aforementioned steady changes in the circulation have given or will give rise to rapid changes in climate in regions marked by large spatial gradients in surface weather (e.g., regions that lie at the interface of dry and raining regions). Additionally, circulation trends that are robust on large spatial scales may be much more difficult to detect on regional spatial scales due to the competing effects of internal climate variability (e.g., Deser et al., 2012a, 2012b).

Summary and the Way Forward

It is difficult to detect *steady* trends in the atmospheric circulation, particularly on regional scales where the trends are superposed on marked internal variability. It is also difficult to detect statistically robust *abrupt* changes in the circulation. Detection of an abrupt climate change requires demonstrating that the system was stationary before and after the change occurred. Furthermore, a seemingly robust abrupt climate shift can readily arise due to the chance superposition of internal and forced climate change. The steady changes in the circulation noted in this section are generally not abrupt, but rather scale with the timescale of the forcing (with the notable exception of the aforementioned Wang et al., 2012b study).

Nevertheless, even relatively steady changes in the atmospheric circulation may prove important for understanding past and future abrupt climate change *if* such changes are coincident with large horizontal gradients in surface climate. Modest and slowly evolving changes in the width of the Hadley Cell could force rapid changes in precipitation in transition regions that lie between the subtropical deserts and tropical rainforests (e.g., the Sahel). Similarly slowly evolving changes in the middle latitude jetstreams could potentially lead to marked changes in surface temperature and precipitation in regions that lie on the flanks of the storm tracks, such as southern Australia. But again, such changes in the circulation will be difficult to detect in the presence of internal climate variability, particularly on regional scales in the extratropics (Deser et al., 2012a, 2012b; Wallace et al., 2013). The ability of steady changes in the circulation to drive abrupt changes in surface climate has not been widely investigated and is a key topic for future research (Box 2.3).

BOX 2.3 COUPLING OF ATMOSPHERIC AND LAND SURFACE AS A CURRENT RESEARCH FRONTIER

The coupling between land surface vegetation and atmosphere could also potentially cause abrupt changes of atmospheric circulation at regional scales. For example, coupled atmosphere-vegetation models suggest that gradual changes in Earth's orbit may lead to the collapse of the Sahara green vegetation and climatic drying (e.g., Bathiany et al., 2012; Claussen et al., 1999; Zeng and Neelin, 2000), although other mechanisms could also be responsible for the latter (e.g., Liu et al., 2006) and in general dynamic vegetation models are in the early stages of development. In a regional model experiment coupled with a simple coupled atmospheric-vegetation model, an abrupt northward jump of the West Africa monsoon circulation can result; when the regional model is initialized with the vegetation/desert border at about 21°N, the low-level westward jetstream over northern Africa and rainfall shift northward and lead to a vegetated central Sahara (Wang and Eltahir, 2000; Patricola and Cook, 2008). However, when the model is initialized by relatively small deviations of the vegetation/desert border from its location today (~10°N), the vegetation distribution tends to remain similar to that found today. Whether such abrupt changes can be reproduced by coupling an atmospheric model with a more comprehensive dynamic vegetation model remains to be seen.

The time-dependent behavior of the atmospheric circulation is generally well monitored by the current observing network of surface stations and radiosondes in the Northern Hemisphere. Circulation measurements in the tropics and Southern Hemisphere are less widespread, and the tropics in particular suffers from a lack of long-term in-situ observations of atmospheric temperatures and pressure. Maintaining and enhancing the current observational network of remotely sensed and in-situ measurements that can be used to infer changes in the atmospheric circulation is essential.

The likelihood of abrupt changes in the atmospheric circulation remains unclear, as does the potential for inducing abrupt climate change in regions of large gradients in surface weather. As such, understanding abrupt changes in—and due to—the atmospheric circulation remains a key topic for future research. Additional investigative work by individual scientists is required in a range of research areas. Interdisciplinary research is needed to assess the importance of changes in the circulation for regions of particular "vulnerability," e.g., in terms of food-security or ecosystems habitat. Model studies are required to assess the mechanisms that drive trends in the circulation, and their amplitudes relative to internal climate variability. Observational studies are required to assess and monitor changes in the observed circulation.

Weather and Climate Extremes

Extreme weather and climate events include heat waves, droughts, floods, hurricanes, blizzards, and other events that occur rarely. In some cases, statistical probability is used to define these extremes, for example, heavy rainfall events or extremely hot or cold temperatures with a 1, 5, or 10 percent occurrence probability. The IPCC SREX report (Seneviratne et al., 2012) defines them as having a 5 percent or 1 percent or even lower chance of occurrence during the same period, and other examples include 10 percent coldest nights, 10 percent warmest daily maximum temperature, 5 percent heaviest rain rate, etc. In other cases, weather and climate extremes are defined by exceeding a threshold that typically results in significant impacts or costs. For example, hurricanes and typhoons defined by wind speeds exceeding specific thresholds are always considered extreme events.

Conditions considered extreme in one location (for example warm temperatures in Barrow, Alaska) may not be extreme in other locations (for example in Miami, Florida). Also, what is normal in one season, a snowfall of three feet in January in New Hampshire, for example, may be extreme at the same location but in another season. Other considerations include the impact of compounding events; for example, some climate extremes, such as droughts or floods, may be the result of an accumulation of moderate weather or climate events. The individual events may not extreme, but the accumulation of them over a relatively short period of time leads to an extreme event, in which case the closely-spaced accumulation of the events is the extreme.

Extreme weather and climate events are among the most deadly and costly natural disasters. For example, tropical cyclone Bhola in 1970 caused about 300,000-500,000 deaths in East Pakistan (Bangladesh today) and West Bengal of India.[3,4] Hurricane Katrina caused more than 1,800 deaths and $96-$125 billion in damages to the Southeast U.S. in 2005. Worldwide, more than 115 million people are affected and more than 9,000 people are killed annually by floods, most of them in Asia (Figure 2.9 or see, for example, the Emergency Events Database[5]). Heat waves contributed to more than 70,000 deaths in Europe in 2003 (e.g., Robine et al., 2008) and more than 730 deaths and thousands of hospitalizations in Chicago in 1995 (Chicago Tribune, July 31, 1995; Centers for Disease Control and Prevention, 1995). Heat waves are one of the largest weather-related sources of mortality in the United States annually.[6]

According to data collected by the National Climate Data Center, there were 134 weather or climate disaster events with losses exceeding $1 billion each in the United States between 1980 and 2011, an average of more than four per year (Table 2.1). Floods, droughts and wildfires—events that appear to be changing in frequency and severity due to climate change—make up about a third of these and slightly more than a third of the dollar damages (adjusted to 2012 dollars). Droughts are particularly costly, comprising about 12 percent of the events by number, but about double that (23.8 percent) by total cost.

Climate Change Is Affecting Extremes

Climate change is expected to shift frequency statistics for weather and climate events, as illustrated in Figure 2.10, in ways that affect the likelihood of extreme events on the tails of the distribution, either the high side ("extremely hot" for example) or the low side ("extremely cold"). Such shifts are already being observed. For example, Hansen et al. (2012) studied temperature anomalies over the past 6 decades and found that while anomalies greater than three standard deviations occurred over about 0.3 percent of the land area in their base period (1951-1980), they now

[3] http://www.wunderground.com/hurricane/deadlyworld.asp.

[4] http://www.nbcnews.com/id/24488385/ns/technology_and_science-science/t/deadliest-storms-history/#.UX1b2KX2Wqx.

[5] http://www.emdat.be/.

[6] http://www.nws.noaa.gov/os/heat/index.shtml.

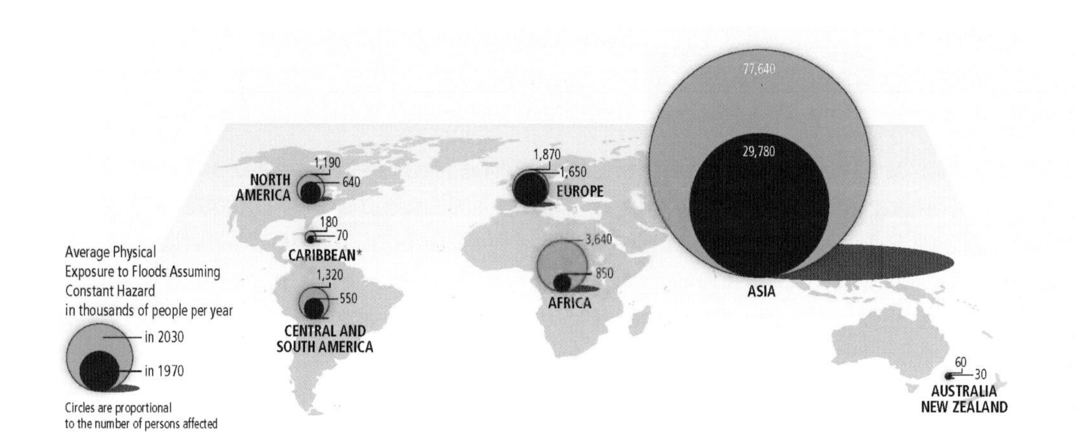

FIGURE 2.9 The projected increase in the number of people (in thousands) exposed to floods in 2030 compared to those in 1970. Only catchments bigger than 1,000k m^2 were included in analysis; therefore, only the largest islands in the Caribbean are covered. Source: IPCC, 2012; Solterra Solutions, 2012.

TABLE 2.1 Billion-dollar weather and climate disasters in the United States from 1980 to 2011 by type. Total damages are in consumer-price-index-adjusted 2012 dollars. Note that the impacts of droughts are difficult to determine precisely, so those figures may be underestimated.

Disaster Type	# Events	% Frequency	CPI-adjusted Damages (billions of dollars)	% Damage
Severe storm	44	32.8	96.1	10.9
Tropical cyclone	31	23.1	417.9	47.4
Flooding	16	11.9	85.1	9.7
Drought	16	11.9	210.1	23.8
Wildfire	11	8.2	22.2	2.5
Winter storm	10	7.5	29.3	3.3
Freeze	6	4.5	20.5	2.3
Total	**134**	**100%**	**881.2**	**100%**

Source: Solterra Solutions, 2012.

(2006–2011) occur on 6-17 percent of the land (Figure 2.11; see also the comment on Hansen et al. [Rhines and Huybers, 2013], and Hansen et al.'s response [Hansen et al., 2013b]). A similar change has been observed in rainfall. While total precipitation in the United States increased by about 7 percent over the past century, the heaviest 1 percent of rain events increased by nearly 20 percent (Bull et al., 2007).

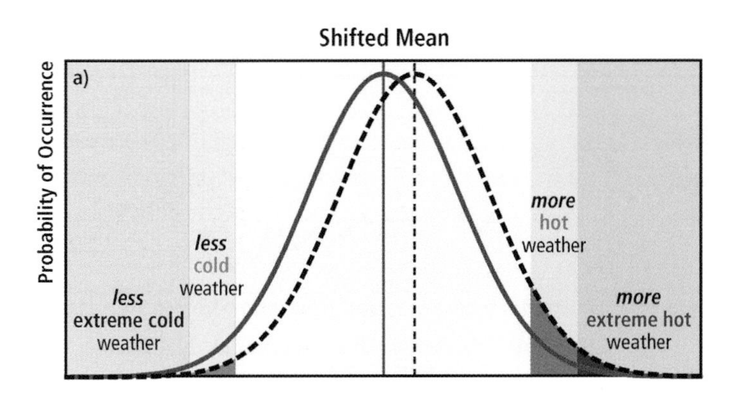

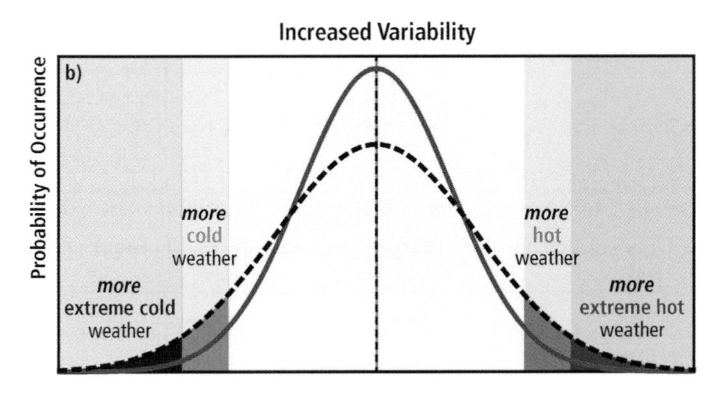

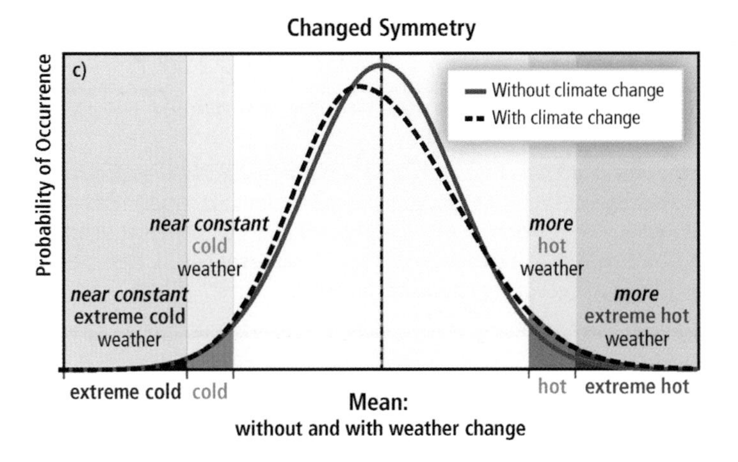

FIGURE 2.10 Potential effects of changes in temperature distribution on extremes: a) effects of a simple shift of the entire distribution toward a warmer climate; b) effects of an increased temperature variability with no shift of the mean; and c) effects of an altered shape of the distribution, in this example an increased asymmetry toward the hotter part of the distribution. SOURCE: Lavell et al., 2012.

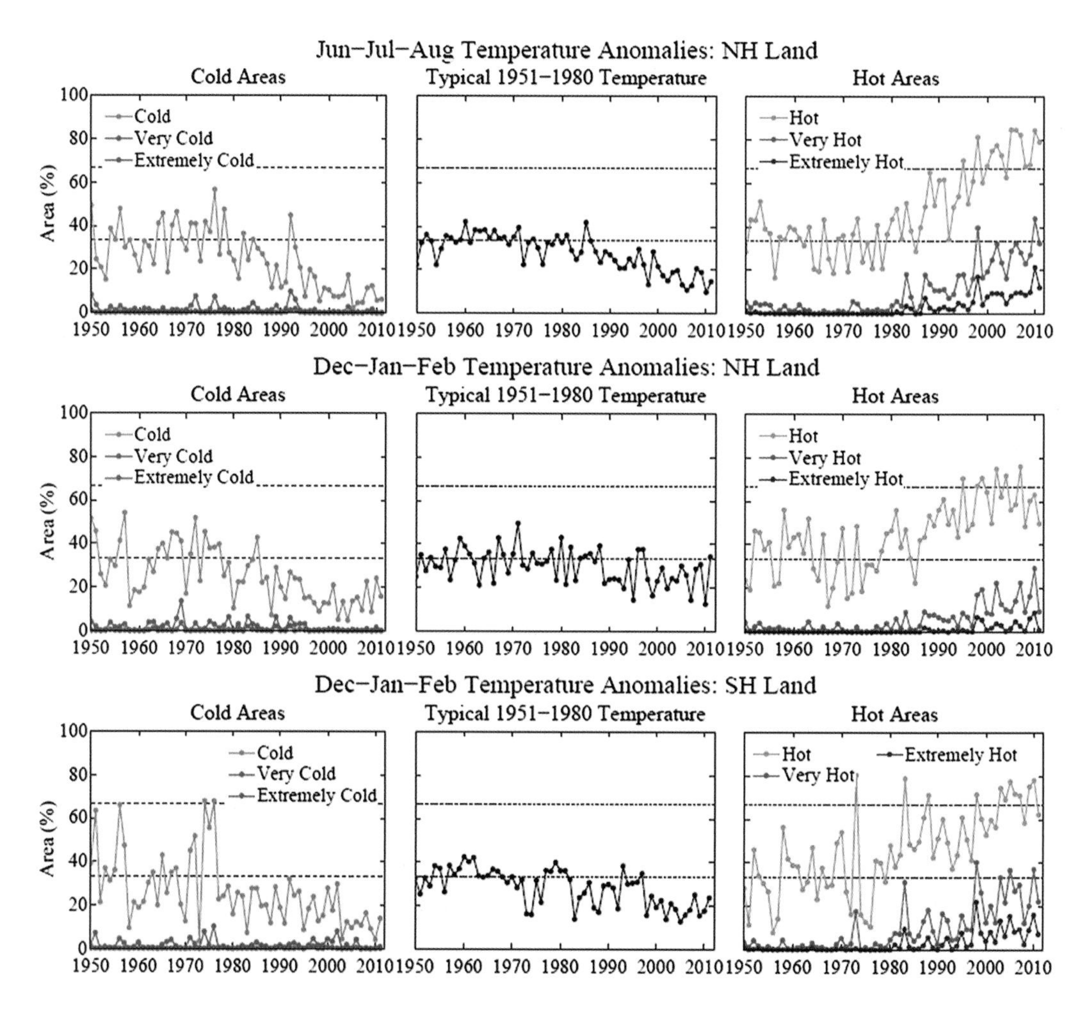

FIGURE 2.11 Area of the world (in percent) covered by temperature anomalies (from Figure 5 in Hansen et al., 2012) in categories defined as hot (+0.43 σ), very hot (+2 σ), and extremely hot (+3 σ), with analogous divisions for cold anomalies. Dashed horizontal lines indicate areas of 33 percent and 67 percent, corresponding to climate dice with two and four sides colored red, respectively. Note: NH=Northern Hemisphere; SH=Southern Hemisphere. Source Hansen et al., 2012.

Climate change may also be affecting other weather and climate extremes, with impacts and trends that vary regionally. The Mediterranean and West Africa are likely experiencing more frequent and severe droughts, while the opposite is the case for central North America and northwest Australia (IPCC, 2013). Longer, hotter, and drier summers have contributed to an increase in the frequency and severity of wildfire in the western United States (Westerling et al., 2006), a trend that is expected to continue as the climate warms (NRC, 2011a). The possibility of climate change leading to more intense tropical cyclone activity, particularly in the North Atlantic, continues to receive significant research attention.

Links Between Extreme Events and Abrupt Change

While extreme events *per se* are not abrupt climate changes as defined in this report, changes in extreme events could lead to abrupt changes in two ways: (1) an abrupt change in a weather or climate extremes regime, for example a sudden shift to persistent drought conditions; or (2) a gradual trend in the frequency or severity of extremes that causes abrupt impacts when societal or ecological thresholds are crossed, as illustrated in Figure 2.10. The potential for abrupt regime shifts was raised in NRC (2002), which highlighted the transitions into and out of the 1930s Dust Bowl as prime examples. Since NRC, 2002 was published, the potential for abrupt impacts associated with gradual changes in extreme events—such as abrupt changes in terrestrial ecosystems due to droughts and storms—has been studied extensively (e.g., Hutyra et al., 2005; Saatchi et al., 2013). Moreover, the vulnerabilities of the complex and globally connected socio-economic system to such extreme events have become more widely recognized (Mileti, 1999; NRC, 2010a; NRC, 2012c) and the impacts of extreme events on societal tipping points have been more clearly appreciated (Lenton et al., 2008; Nel and Righarts, 2008).

Most extreme events are temporary and their impacts are reversible. For example, the US Great Plains recovered from the severe drought in the 1930s when rainfall returned to normal and land management practices were improved and maintained. However, even temporary extreme climatic events can trigger abrupt and irreversible changes when their impacts exceed the threshold or resilience of the ecosystems. For example, the 1950s drought in New Mexico abruptly shifted the ecotone between semiarid ponderosa pine forest and piñon–juniper woodland (Allen and Breshears, 1998).

In addition to a changing climate causing changes in extreme events, extreme events themselves can accelerate abrupt changes in other parts of the climate and Earth system. For example, extreme transient sea-level rise due to tropical or extratropical storm surge can cause abrupt increases of flood risk (Nicholls et al., 2007), putting many coastal regions at risk for loss of human life before gradual sea-level rise inundates the region. Extreme warm temperatures in summer can greatly increase the risks of mega-fires in temperate forests, boreal forests, and savanna ecosystems, leading to abrupt changes in species dominance and vegetation type, regional water yield and quality, and carbon emission (e.g., Adams, 2013), before the gradual increase of surface temperature crosses the threshold for abrupt ecosystem collapse (more discussion in the section on Ecosystem Collapse and Rapid State Change below).

Extreme events could lead to a tipping point in regional politics or social stability. In Africa, extreme droughts and high temperatures have been linked to an increase of risk of civil conflict and large-scale humanitarian crisis in Africa (Burke et al., 2009; Hsiang et al., 2011, 2013; Miguel et al., 2004; O'Loughlin et al., 2012; Zhang et al., 2011). Generally, extreme climate events alone do not cause conflict. However, they may act as an accelerant of instability or conflict, placing a burden to respond on civilian institutions and militaries around the world (NRC, 2012b). For example, the devastating tropical cyclone Bhola in 1970 heightened the dissatisfaction with the ruling government and strengthened the Bangladesh separatist movement. This led eventually to civil war and independence of Bangladesh in 1971 (Kolmannskog, 2008; NRC, 2012b). Historically, extreme climate events such as decadal mega-droughts may have triggered the collapse of civilizations, such as the Maya (Hodell et al., 1995; Kennett et al., 2012) or large scale civil unrest that ended the Ming dynasty (Shen et al., 2007). More extensive review of the extreme climate impacts is provided by the recent IPCC-SREX report (Field et al., 2012) and the 2012 NRC report on *Climate and Social Stress* (NRC, 2012b).

Summary and the Way Forward

The connection between extreme climate and related abrupt climate change is poorly understood, given the relatively poor understanding of both extreme climate events and abrupt changes. A number of reasons exist for this. First, because extreme climate phenomena represent rare events and modern climate records made by instruments are short, the modern record may capture only a few instances of these extreme events. Second, the statistical tools to which most climate researchers are accustomed are not applicable to this highly non-linear problem. Third, lack of quantitative understanding of the thresholds that trigger abrupt changes and causes of extreme climate events has limited our ability to provide process-based assessments of the risk of abrupt changes. Extreme events and the resultant abrupt changes are more likely unpredictable based on statistical models (Ditlevsen and Johnsen, 2010; Hastings and Wysham, 2010). Yet, it is prudent to assess the societal vulnerability and develop no-regret mitigation policies for high-impact extreme events related to abrupt changes (NRC, 2012b). In this case, risk assessment based on a fundamental understanding of the climate dynamics may become a major tool for developing scenarios for stress tests for the global and regional responding systems regarding their ability to manage potentially disruptive extreme and abrupt climate changes.

Coupled global climate models, such as those that participated in the Paleoclimate Modeling Intercomparison Project (PMIP) (Jansen et al., 2007), in combination with improved paleo-climate records have led to better appreciation of the extent of extreme events that have occurred in the past (e.g., Cook et al., 2010b). Finally, possible early warning for some of the abrupt climatic changes has begun to be explored. However, the understanding of the connections between climate change, climate and weather extremes, and abrupt change is still limited (e.g., Seneviratne et al., 2012). This report uses examples provided in the literature to illustrate the potential connections between extreme climate events and abrupt climate change, and highlights the need for a focused research effort to explore these climate events with high societal consequences but low probability.

In recent years, researchers, mainly in Europe, have begun to explore the feasibility of detecting early warning signs of abrupt climate changes. These studies have shown that an early warning signal may be detectable if an abrupt change is driven by gradual forcing and preceded by the critical slow down, increased variances, and skewness (e.g., Held and Kleinen, 2004; Livina and Lenton, 2007). However, extreme events are mainly a result of natural climate variability, making it hard to detect early warning signals within an otherwise noisy time series. The resultant abrupt changes are generally intrinsically unpredictable (e.g., Scheffer et al., 2009; Lenton, 2011). Given such a challenge, risk assessment would depend more on our predictive understanding and process-based probabilistic prediction than on statistical early warning signs of approaching a tipping point. This is central to the ability to improve the quantitative understanding of the thresholds that can trigger abrupt changes and the probability distribution changes of the extreme climate events with the slow varying climate states and forcings that can be monitored.

ABRUPT CHANGES AT HIGH LATITUDES

Potential Climate Surprises Due to High-Latitude Methane and Carbon Cycles

Interest in high-latitude methane and carbon cycles is motivated by the existence of very large stores of carbon (C), in potentially labile reservoirs of soil organic carbon in permafrost (frozen) soils and in methane-containing ices called methane hydrate or clathrate, especially offshore in ocean marginal sediments. Owing to their sheer size, these carbon stocks have potential to massively impact the Earth's climate, should they somehow be released to the atmosphere. An abrupt release of methane (CH_4) is particularly worrisome as it is many times more potent as a greenhouse gas than carbon dioxide (CO_2) over short time scales. Furthermore, methane is oxidized to CO_2 in the atmosphere representing another CO_2 pathway from the biosphere to the atmosphere in addition to direct release of CO_2 from aerobic decomposition of carbon-rich soils.

Permafrost

Stocks Frozen northern soils contain enough carbon to drive a powerful carbon cycle feedback to a warming climate (Schuur et al., 2008). These stocks across large areas of Siberia comprise mainly yedoma (an ice-rich, loess-like deposit averaging ~25 m deep [Zimov et al., 2006b]), peatlands (i.e., histels and gelisols), and river delta deposits. Published estimates of permafrost soil carbon have tended to increase over time, as more field datasets are incorporated and deposits deeper than 1 m depth are considered. Estimates of the total soil-carbon stock in permafrost in the Arctic range from 1,700–1,850 Gt C (Gt C = gigatons of carbon; Tarnocai et al., 2009; Zimov et al., 2006a; McGuire et al., 2009). Figure 2.12 summarizes information on known stocks of high-latitude carbon.

To put the Arctic soil carbon reservoir into perspective, the carbon it contains exceeds current estimates of the total carbon content of all living vegetation on Earth (approximately 650 Gt C), the atmosphere (730 Gt C, up from ~360 Gt C during the last ice age and 560 Gt C prior to industrialization, Denman et al., 2007), proved reserves of recoverable conventional oil and coal (about 145 Gt C and 632 Gt C, respectively), and even approaches geological estimates of all fossil fuels contained within the Earth (~1,500 – 5,000 Gt C). It represents more than two and a half centuries of our current rate of carbon release through fossil fuel burning and the production of cement (nearly 9 Gt C per year, Friedlingstein et al., 2010).

These vast deposits exist largely because microbial breakdown of organic soil carbon is generally low in cold climates, and virtually halted when frozen in permafrost. Despite slow rates of plant growth in the Arctic and sub-Arctic latitudes, massive deposits of peat have accumulated there since the last glacial maximum (Smith et al., 2004; MacDonald et al., 2006).

Potential response to a warming climate Permafrost soils in the Arctic have been thawing for centuries, reflecting the rise of temperatures since the last glacial maximum (~21 kyr ago) and the Little Ice Age (1350-1750). However, this Holocene thawing

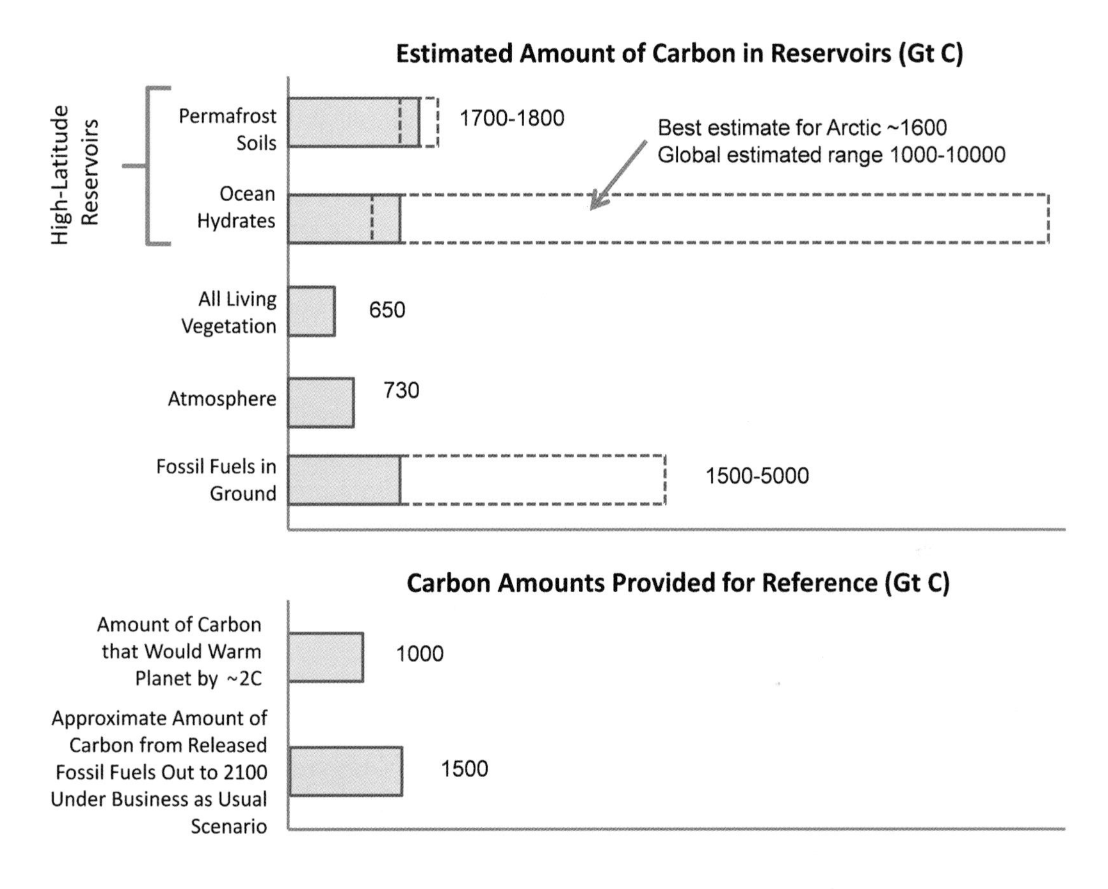

FIGURE 2.12 Top: Approximate inventories of carbon in various reservoirs (see text for references). Bottom: Provided for reference, estimated amount of carbon that would warm the planet approximately 2°C (Allen et al., 2009; uncertainty estimate in this value discussed in this reference) and estimated total amount of carbon to be released by the year 2100 under business-as-usual scenarios (IPCC, 2007c).

has accelerated in recent decades, and can be attributed to human-induced warming (Lemke et al., 2007). Under business-as-usual climate forcing scenarios, much of the upper permafrost is projected to thaw within a time scale of about a century (Camill, 2005, Lawrence and Slater, 2005). Exactly how this will proceed is uncertain. The rate of carbon degradation increases nonlinearly with temperatures above the freezing point of water. Furthermore, the spatial pattern of this degradation is spatially heterogeneous owing to small-scale geomorphic processes such as thermokarsting and slumping from ice-wedge melting (Jorgenson et al., 2006).

Wildfires have been spreading into some permafrost regions as local climatic changes promote increasingly dry conditions. (Lynch and Wu, 2000; Mack et al., 2011; Schuur et al., 2013). Charcoal records cored from 14 lakes in the Alaska interior suggest that recent fires burning there are unprecedented over the past 10,000 years (Kelly et al., 2013). The interaction of boreal fires with overall climate forcing is complex, because carbon release from burned plant material and soil carbon is at least partially countered by increased albedo of the fire scar (Goetz et al., 2007; Randerson et al., 2006). If the fire is sufficiently severe and burns the surface organic layer, heat flow through the active layer increases into the permafrost, and the active layer can increase to a thickness that does not completely refreeze the following winter. This results in formation of a talik, an unfrozen layer below the seasonally frozen soil and above the permafrost (Yoshikawa et al., 2002). Little is known about the potential of such burning to thaw and release stored carbon faster than would occur solely from broader, landscape-scale decomposition, but the magnitude of carbon loss due to fire is significant and potentially offsets the increased carbon sequestration through Arctic greening (Mack et al., 2011). In sum, this known mechanism for rapid, unexpected carbon release demands further research.

The chemical fate of the decomposing carbon (i.e., methane versus CO_2 emission) depends primarily on the availability of oxygen, which is controlled in these settings by how wet the soil is. Dry, well-aerated soils oxidize the carbon to produce CO_2. Wet soils tend to be anoxic, leaving anaerobic fermentation as the degradation pathway. The maximum methane yield fraction is about 50 percent. However, methane can be oxidized to CO_2 in the soil column (Reeburgh, 2007), so the methane fraction of the net carbon emissions to the atmosphere can be, and usually is, much lower than this.

Projecting the future water balance and moisture state of Arctic soils—and thus the ratio of CO_2 to CH_4 production—contributes the largest uncertainty in forecasting methane emissions from Arctic land surfaces. Because present-day permafrost landscapes generally support a greater abundance of lakes and wetlands than do thawed landscapes (Smith et al., 2005, 2007), a complete disappearance of permafrost would suggest an ultimately drier land surface and thus reduced methane production. However, such a transformation would require centuries to millennia, and numerous studies have shown that the initial stages of permafrost degradation lead to paludification (lake formation) of the land surface and increased methane emissions (Skre et al., 2002). Furthermore carbon-rich peatlands, when thawed, retain large volumes of liquid water and may have surfaces even moister than their frozen peatlands (Smith et al., 2012).

Climate-induced permafrost thaw is amenable to numerical modeling, because good theoretical frameworks of how heat propagates from the air-ground interface into the subsurface have been developed. While such models lack adequate observational datasets of subsurface soil properties and/or geology, it is clear that the time scale for deep permafrost thaw is measured in centuries, not years. Furthermore, unlike methane hydrates (see below), the very large stocks of permafrost soil carbon (i.e., the 1,672 Gt C of Tarnocai et al., 2009) must first undergo anaerobic microbial fermentation to produce methane, itself a gradual decomposition process. There are no currently proposed mechanisms that could liberate a climatically significant amount of methane or CO_2 from frozen permafrost soils within an abrupt time scale of a few years, and it appears gradual increases in carbon release from warming soils can be at least partially offset, owing to rising vegetation net primary productivity (Beilman et al., 2009). Over a time scale of decades, however, a possible self-sustaining decomposition of Yedoma could occur before the end of this century (Khvorostyanov et al., 2008a, 2008b, 2008c). A related idea is the possibility of rising soil temperatures triggering a "compost bomb instability" (Wieczorek et al., 2011)—possibly including combustion—and a prime example of a rate-dependent tipping point (Ashwin et al., 2012). Such possibilities would represent a rapid breakdown of the Arctic's very large soil carbon stocks and warrant further research. Even absent an abrupt or catastrophic mobilization of CO_2 or methane from permafrost carbon stocks, it is important to recognize that Arctic emissions of these critical greenhouse gases are projected to increase gradually for many decades to centuries, thus helping to drive the global climate system more quickly towards other abrupt thresholds examined in this report.

Methane Hydrates in the Ocean

Stocks Under conditions of high pressure, high methane concentration, and low temperature, water and methane can combine to form icy solids known as methane hydrates or clathrates in ocean sediments. The methane derives from biological or thermal degradation of organic matter originally deposited on the sea floor. Although the overall rate of methane production in ocean sediments is fairly slow, over millions of years, substantial reservoirs of methane hydrate have accumulated in the world's ocean margins.

Throughout most of the world ocean, a water depth of about 700 m is required for hydrate stability. In the Arctic, due to colder-than-average water temperatures, only about 200 m of water depth is required, which increases the vulnerability of those methane hydrates to a warming Arctic Ocean. The Arctic is also a focus of concern because of the wide expanse of continental shelf (25 percent of the world's total), much of which is still frozen owing to its exposure to the frigid atmosphere during lowered sea levels of the last glacial maximum (see above).

The inventory of methane in ocean margin sediments is large but not well constrained, with a generally agreed upon range of 1,000-10,000 Gt C (Archer, 2007; Boswell, 2007; Boswell et al., 2012). One inventory places the total Arctic Ocean hydrates at about 1,600 Gt C by extrapolation of an estimate from Shakhova et al. (2010a) to the entire Arctic shelf region (Isaksen et al., 2011) (see Figure 2.12). The geothermal increase in temperature with depth in the sediment column restricts methane hydrate to within a few hundred meters thickness near the upper surface of the sediments (e.g., Davie and Buffett, 2001). Beneath this stability zone, a layer rich in methane bubbles is often seen in seismic reflection data, called a "bottom simulating reflector." The areal extent of methane-rich sediments is fairly well known from seismic observations of this feature, but uncertainty in the concentration of methane in those sediments is very large, thus resulting in the large uncertainty in the global inventory of ocean-floor methane.

Potential response to a warming climate Climate change has the potential to impact ocean methane hydrate deposits through changes in ocean water temperature near the sea bed, or variations in pressure associated with changing sea level. Of the two, temperature changes are thought to be most important, both during the last deglaciation (Mienert et al., 2005) and also in the future. Warming bottom waters in deeper parts of the ocean, where surface sediment is much colder than freezing and the hydrate stability zone is relatively thick, would not thaw hydrates near the sediment surface, but downward heat diffusion into the sediment column would thin the stability zone from below, causing basal hydrates to decompose, releasing gaseous methane. The time scale for this mechanism of hydrate thawing is on the order of centuries to millennia, limited by the rate of anthropogenic heat diffusion into the deep ocean and sediment column. Even on the Siberian continental margin, where water temperatures are colder than the global average, and where the sediment column retains the cold imprint from its exposure to the atmosphere during the last glacial time 20,000 years ago, any methane hydrate must be buried under at least 200 m of water or sediment. Bottom waters at depths of 50 or 100 m might warm relatively quickly with a collapse in sea ice cover, but it would take centuries for that heat to diffuse through the 100-150 m of sediment column to the hydrate stability zone. Thus the release of 50 Gt C from the Siberian continental shelf in 10 years as postulated by Whiteman et al. (2013) is unlikely.

The proportion of this gas production that will reach the atmosphere as CH_4 is likely to be small. To reach the atmosphere, the CH_4 would have to avoid oxidization within the sediment column (a chemical trap) and re-freezing within the stability zone shallower in the sediment column (a cold trap). However, the hydrate stability zone thickness decreases to zero near the top of its depth range in the ocean, and an increase in water column temperature there could eliminate the stability zone entirely, potentially providing an easier pathway for methane to reach the sea floor. Episodic and explosive escapes of gaseous methane from the sediment column have been documented by kilometer-scale "wipeout zones" in seismic images (Riedel et al., 2002), and pockmarks on the sea floor, called eruption craters (Hill et al., 2004). However, the processes responsible for these observations are too poorly understood to predict what fraction of deeper CH_4 might be released through them.

Most of the methane gas that emerges from the sea floor dissolves in the water column and oxidizes to CO_2 instead of reaching the atmosphere. Bubble plumes tend to dissolve on a height scale of tens of meters (Rehder et al., 2002; Kessler et al., 2011), although larger plumes, consisting of larger bubbles, do rise farther. However, even in the cold Arctic Ocean, methane hydrate is only stable below about 200 m water depth, making for an inefficient pathway to the atmosphere at best. The highest oceanic methane fluxes to the atmosphere in the Arctic are probably in the coastal zone, associated with erosion of coastal permafrost (Shakhova et al., 2010b). In this region (and also in terrestrial lakes) the methane flux to the atmosphere is strongly impacted by ice formation on the water surface (Walter et al., 2007), providing another mechanism for climate feedback (He et al., 2013).

Another, more abrupt way to transfer methane hydrate from the sediment column to the atmosphere is by way of a submarine landslide. Methane hydrate floats in seawater just as water ice floats, and it also has greater potential to reach the atmosphere than methane bubbles (Brewer et al., 2002). The largest known submarine landslide (called Storegga) occurred ~8000 years ago, as documented in sediment deposits off Norway (Mienert et al., 2005). The volume of sliding material multiplied by a reasonable hydrate fraction in the pore space yields a possible methane source of about 1 Gt C. The climatic impact of this quantity of methane would be comparable to that of a volcanic eruption (although warming rather than cooling). As such it would have a significant climate impact, but one that is likely to be smaller than that of the anthropogenic CO_2 rise (Archer, 2007).

Over time scales of centuries and millennia, the ocean hydrate pool has the potential to be a significant amplifier of the anthropogenic fossil fuel carbon release. Because the chemistry of the ocean equilibrates with that of the atmosphere (on time scales of decades to centuries), methane oxidized to CO_2 in the water column will eventually increase the atmospheric CO_2 burden (Archer and Buffett, 2005). As with decomposing permafrost soils, such release of carbon from the ocean hydrate pool would represent a change to the Earth's climate system that is irreversible over centuries to millennia.

Modeling the response of ocean hydrates to climate change is in its infancy. The largest uncertainty is the concentration of methane hydrate, especially in the shallow sediment column near the sediment water interface. Coupled atmosphere-ocean climate models can be used to simulate the thermal response of the ocean water column to climate change with a moderate degree of uncertainty and the subsequent penetration of heat into the sediment column. The response of an assumed column inventory of hydrate to warming can be simulated (Lamarque, 2008; Reagan and Moridis, 2009; Reagan et al., 2011), but the results depend strongly on the assumed hydrate concentrations. Another approach is to "grow" the sediment column through geologic time to obtain an initial condition for a climate change perturbation scenario (Archer et al., 2012), but uncertainties in various model parameters, such as the methane production rate and the fate of bubbles in the sediment column, prevent a well-constrained model forecast of the methane hydrate response to climate warming.

In summary, the ocean methane hydrate pool has strong potential to amplify the human CO_2 release from fossil fuel combustion over times scales of decades to centuries. While anthropogenic warming should accelerate the thawing of offshore permafrost via warming of Arctic Ocean shelf waters, this impact should be considered additive to a broader thawing trend that has been underway for thousands of years.

Impacts of Arctic Methane on Global Climate

Although attention is often focused on methane when considering a potential Arctic carbon release, because methane is a short-lived gas in the atmosphere (CH_4 oxidizes to CO_2 within about a decade), ultimately a methane problem is a CO_2 problem. It does matter how rapidly methane is released, and the impacts of a spike versus chronic emissions are discussed in Box 2.4. As methane emissions from permafrost degradation will also be accompanied by larger fluxes of CO_2, Arctic carbon stores clearly have the potential to be a significant amplifier to the human release of carbon. The impact of Arctic carbon release on carbon policy thus should be considered. If anthropogenic carbon emissions are limited by law and/or economic means, yet CO_2 and CH_4 levels do not respond as expected because Arctic emissions are increasing, this could impact the willingness of countries to engage in limiting human carbon emissions.

Speculations about potential methane releases in the Arctic have ranged up to about 75 Gt C from the land (Isaksen et al., 2011) and 50 Gt C from the ocean (Shakhova et al., 2010a). A release of 50 Gt C methane from the Arctic to the atmosphere over 100 years would increase Arctic CH_4 emissions by about a factor of 25, and would make the present-day permafrost area about two times more productive of CH_4 on average as comes from wetlands today. Postulating such a methane release over a more abrupt 10-year time scale, the emission rates from present-day permafrost would have to exceed that from wetlands by a seemingly implausible factor of 20, supporting a longer century timescale for this process, and making methane emission from polar regions an unlikely candidate for a tipping point in the climate system. Nonetheless, as can be seen in Box 2.4, releasing 50 Gt C of methane over 100 years would have a significant impact on Earth's climate. The atmospheric CH_4 concentration would roughly quadruple, with a resulting total radiative forcing from CH_4 of about 3 Watts/m^2. The magnitude of this forcing is comparable to that from doubling the atmospheric CO_2 concentration, but the impact of the methane forcing would be strongly attenuated by its short duration (see Box 2.4).

The impact of the Arctic as a source of natural methane and CO_2 can be monitored by measuring the concentrations of these gases in samples from around the world (in combination with models), as is being done by the NOAA Carbon Cycle Greenhouse Gas program[7], although maintaining these networks in an era of budget cuts is an area of concern. As concluded above, an increase in Arctic CH_4 emissions of more than a factor of 10 is required before it would begin to have a significant impact on Earth's climate in the short term. Such a strong acceleration of methane degassing from the Arctic would result in measurably higher concentrations of methane in the high northern latitudes. Another avenue for monitoring is satellite measurements of column inventories of the gases, which provide much more detailed spatial coverage but no vertical resolution, in which air masses at different altitudes may carry gases that originated from different parts of the Earth's surface. Measurements from aircraft, manned and unmanned, are the third potential monitoring approach, providing vertical resolution of the concentrations, which gives much tighter constraint on local-source fluxes.

Summary and the Way Forward

Arctic carbon stores are poised to play a significant amplifying role in the century-timescale buildup of CO_2 and methane in the atmosphere, but are unlikely to do so abruptly, on a time scale of one or a few decades. This conclusion is based on immature science, however, and a truly sparse monitoring capability. Marine hydrates are poorly mapped, and the possibility that they could even become a targeted fossil-fuel resource for future economic development cannot be dismissed. Basic scientific research is required to assess the long-term stability of currently frozen Arctic and sub-

[7] http://www.esrl.noaa.gov/gmd/ccgg/flask.html.

BOX 2.4 ARCTIC CARBON METHANE RELEASE: SPIKE OR CHRONIC?

The response of atmospheric concentration to a methane release depends on whether the release time scale is shorter or longer than the atmospheric lifetime of methane. An instantaneous release, for example, would cause the atmospheric methane concentration to spike immediately, then decay back toward the unperturbed value on a time scale of approximately one decade.

The climatic impact of a spike of methane would be shaped by the long time scale of the Earth's temperature response to radiative (greenhouse gas) forcing, which in turn is set by the absorption of heat energy by the deep ocean on time scales of centuries to a millennium. The impact of the ocean uptake of heat would be to strongly attenuate the short-term climatic impact of such a spike in radiative forcing (see Figure). However, the ocean uptake of heat would also act to "bank" the heat, accumulating it through the spike period, prolonging the recovery of surface temperature beyond the demise of the methane spike itself. The spike therefore serves as a source of long-term ocean thermal pollution, which would be added to that from the anthropogenic atmospheric CO_2 rise.

If, on the other hand, a methane release to the atmosphere continues for much longer than the methane lifetime, the concentration of methane in the atmosphere will rise to a new steady-state value. In general, the concentration of methane in the atmosphere is expected to scale roughly linearly with the global emission flux. The potential increase in chronic methane emissions from the Arctic must therefore be evaluated in the context of global methane emissions. These fluxes are compared in the Table below. Present-day methane emissions from the Arctic are much smaller than natural emissions, mostly from tropical wetlands, and human emissions (Denman et al., 2007). For this reason it would require a very large, prolonged relative increase in Arctic sources to significantly affect Earth's climate.

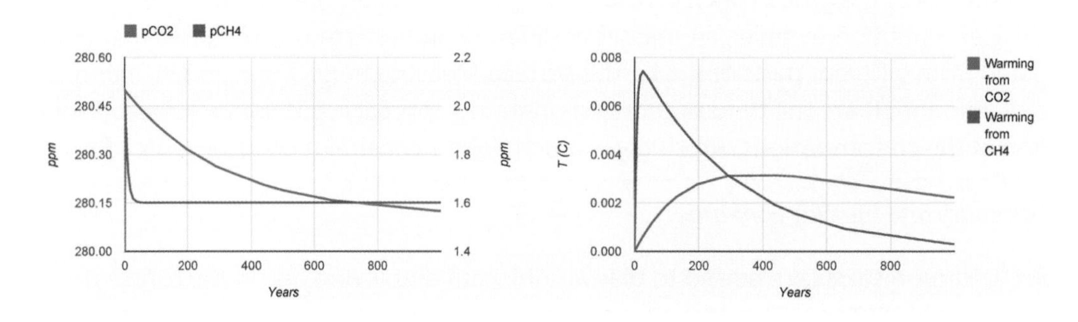

FIGURE Atmospheric chemistry and climatic impact of an abrupt 1 GtC methane release. (Left) methane concentration spikes after the release at the beginning of the simulation, then is oxidized to CO_2. (Right) The temperature as driven by CH_4 radiative forcing increases strongly during the methane spike, then subsides following the time scale of planetary (oceanic) cooling. The temperature change due to the oxidized CO_2 grows but also subsides more slowly than that from CH_4, due to the weaker greenhouse forcing but longer atmospheric lifetime of the CO_2. From http://forecast.uchicago.edu/Projects/slugulator.html, based on data from Archer et al., 1997 and Schmidt and Shindell, 2003.

BOX 2.4 CONTINUED

TABLE Summary of methane release scenarios compared with present-day methane fluxes and the radiative impact of business-as-usual CO_2 rise.

Scenario	CH_4 emission rate, Gt C	Arctic Increase factor relative to today	Arctic CH_4 flux/ Wetland flux, per m^2 area	CH_4 Conc. in Steady State	Radiative Forcing
Natural	0.15				
Anthropogenic	0.25				
Arctic lakes (Walter 2007)	0.02				
50 Gt C over 10 years	5	250	20	20 ppm	5 Watts/m^2
50 Gt C over 100 years	0.5	25	2	6 ppm	3 Watts/m^2
Business-as-usual CO_2 in 2000 (500 Gt C released overall)					1.5 Watts/m^2
Business-as-usual CO_2 in 2100 (~1500 Gt C)					6 Watts/m^2

SOURCES: IPCC AR4 Ch 7, IPCC, 2000.

Arctic soil stocks, their future hydrologic status (i.e., moister or drier) that will largely drive their methane emissions, and the possibility of increasing methane gas bubble ebullition from currently frozen marine and terrestrial sediments as their temperatures rise.

If permafrost soils begin to release climatically significant amounts of methane, it should be detectable through monitoring atmospheric concentrations of methane using a network of monitoring stations around the world, but the current network is too sparse. Satellite observations of atmospheric chemistry would provide another means of detecting an Arctic methane feedback. Both types of sampling also provide constraint on sources and sinks of CO_2, which play a significant part of the potential climate impact of the Arctic. It is therefore vital that the flask sample and satellite atmospheric chemistry monitoring efforts be continued and expanded.

While it is not possible to directly observe subsurface permafrost state from satellite remote sensing, surface freeze-thaw status is readily observed in radar scatterometer images, making this technology one of the best ways to infer frozen vs. thawed ground conditions over large, remote geographic areas. Below the ground surface, *in-situ* methods offer the most direct and effective way to monitor the state of permafrost health, through ongoing temperature measurements in boreholes. However, both the number and geographic extent of long-term borehole observing sites is small. Approximately 200 boreholes of varying depths have been identified for permafrost monitoring by International Permafrost Association (IPA) Global Terrestrial Network for Permafrost[8] (GTN-P), which includes monitoring activities of the Geological Survey of Canada. The geographic coverage represented by these sites is greatest (by far) in Alaska, especially when shallow surface (<10m depth) boreholes are considered. Coverage is especially sparse in the continental interiors of Canada and Russia, and most notably so in the vast James Bay and West Siberian lowlands, as they contain very large stocks of frozen soil carbon in the form of peatland soils that have accumulated since the last glacial maximum.

A second key component for permafrost monitoring is measurements of active-layer depth (the thickness of seasonally thawed soil, measured downward from the soil surface). The dominant monitoring program in this respect is the IPA Circumpolar Active Layer Monitoring Network (CALM), which since its inception in 1991 has developed a network of more than two hundred monitoring sites in fifteen countries, mostly in the Arctic and sub-Arctic.[9] Similar to the borehole monitoring sites (and indeed, often coincident with them), the geographic coverage of active-layer monitoring sites is

[8] http://www.gtnp.org.

[9] http://www.gwu.edu/~calm/.

sparse, with greatest concentration in Alaska and a glaring absence of sites in carbon-rich permafrost soils of interior Canada and Russia. Given that thawing permafrost also affects buildings, roads and other infrastructure, and thus society has a direct stake in its progression, there may also be an opportunity for citizen science in establishing additional permafrost monitoring sites.

Sea Ice

The Arctic Ocean has historically been largely covered in sea ice, which changes considerably with season and plays an integral role in the global climate system. Arctic sea ice has undergone rapid change since satellite records began in 1978. Significant decreases in sea ice have occurred during all months, but the most notable ice losses have occurred in summer. The linear trend in September sea ice from 1979-2012 was a loss of 13 percent per decade relative to the 1979-2000 mean (Fetterer et al., 2012; Stroeve et al., 2012a). The long-term decreases in summer sea ice are superposed on extreme record minima in 2007 and even less in 2012 (Stroeve et al., 2008),[10] with the record low in Arctic sea ice extent on September 16, 2012 of only approximately 3.4 million square km.[11] This September 2012 sea ice extent minimum was only 49 percent of its 1979-2000 mean. Thus, rapid Arctic sea ice loss is underway (Figures 2.13 and 2.14), and given the definitions used in this report, the changes already experienced qualify as an abrupt climate change. Projections from climate models suggest that ice loss will continue in the future, with a possibility of September ice-free conditions later this century (e.g., Stroeve et al., 2012b; Massonnet et al., 2012).

Scientific Understanding of Sea Ice Loss

Past climate models, as judged by the performance of the majority of Coupled Model Intercomparison Project 3 (CMIP3) simulations used in the IPCC Fourth Assessment Report, underestimated the observed linear trend in Arctic sea ice loss (Stroeve et al., 2007). The newer CMIP5 simulations that are being used in the upcoming IPCC Fifth Assessment Report are in better agreement with the observed sea ice loss (Stroeve et al., 2012a; Massonnet et al., 2012), but the reasons for the differences in sea ice trends between the CMIP3 and CMIP5 models remain unclear. They may result from more tuning of sea ice conditions within the models, improved model parameterizations and processes, or some combination of the two.

[10] Also see http://nsidc.org/arcticseaicenews/2012/09/.
[11] http://nsidc.org/arcticseaicenews/2012/09/.

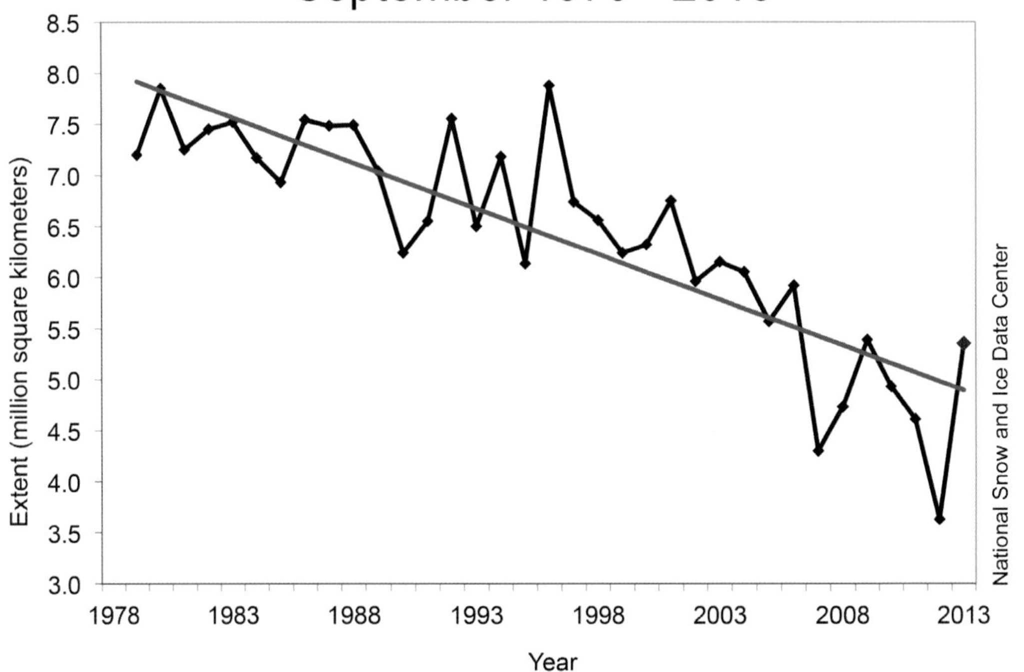

FIGURE 2.13 The time series of September Arctic sea ice extent from 1979-2013. SOURCE: National Snow and Ice Data Center, http://nsidc.org/arcticseaicenews.

A series of extremely low September sea ice conditions during the last decade, including the unprecedented declines in 2007 and 2012, suggests a recent acceleration in the long-term Arctic sea ice loss (e.g., Stroeve et al., 2012b). This bears resemblance to the so-called Rapid Ice Loss Events simulated in a number of climate models (Holland et al., 2006). These simulated events result when anthropogenic change is reinforced by natural variations. They appear to be triggered by increases in ocean heat transport from the North Atlantic to the Arctic and are amplified by the ice-albedo feedback. In the most dramatic of the simulated events, the September ice pack undergoes a 4 million square km loss (about 60 percent of the 1979-2000 ice cover) in only a decade, leading to near ice-free September conditions by 2040.

The rapid nature of observed and predicted changes in the Arctic suggests that Arctic sea ice could possibly undergo nonlinear threshold behavior as it retreats. Arctic climate change is strongly influenced by the surface albedo feedback, which acts to

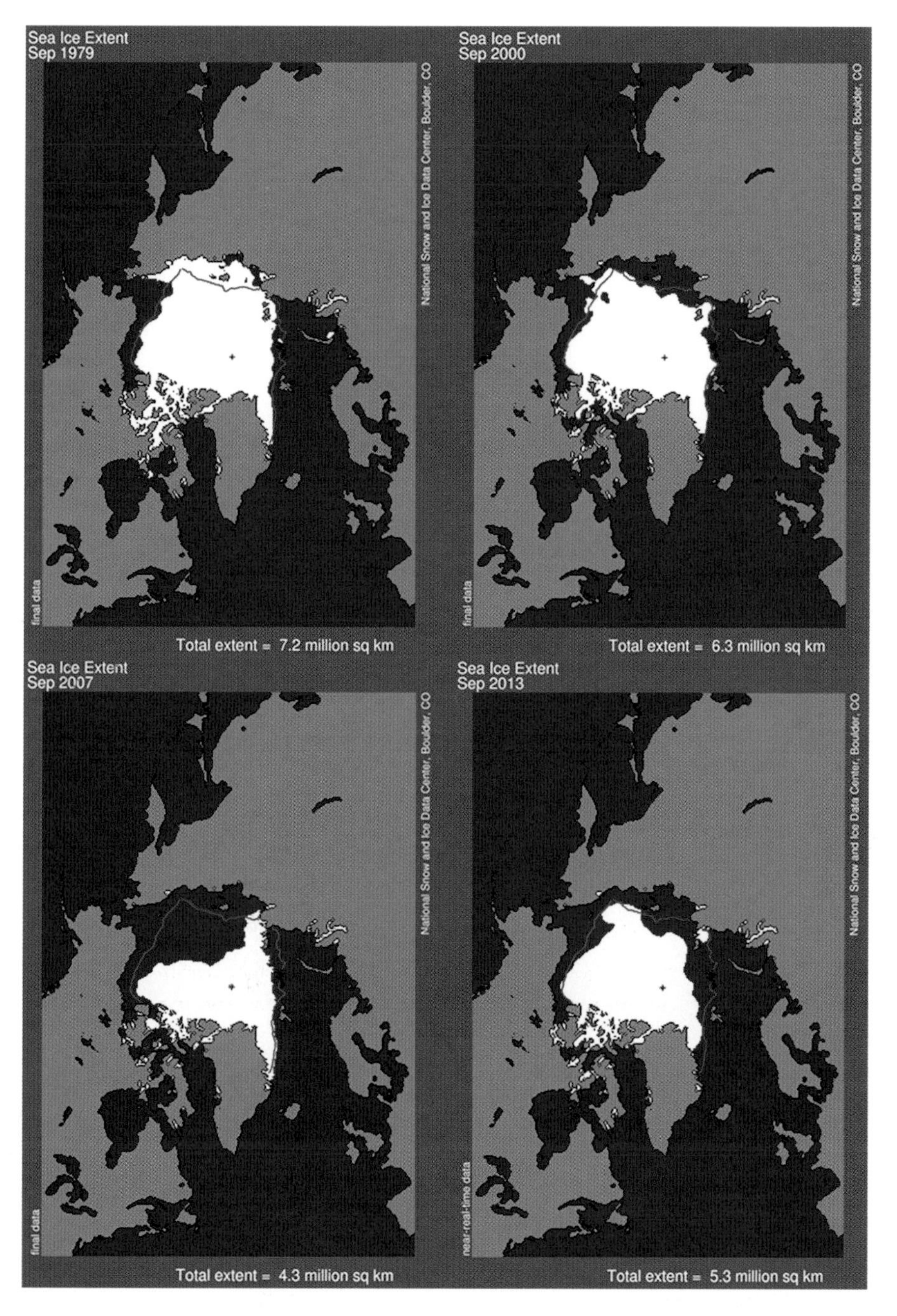

FIGURE 2.14 Extent of Arctic sea ice in September 1979, 2000, 2007, and 2013. The magenta line shows the 1981 to 2010 median extent for September. The black cross indicates the geographic North Pole. Source: National Snow and Ice Data Center, http://nsidc.org/data/seaice_index/.

amplify warming as the reflective ice cover melts and exposes dark open ocean (see Figure 2.15). It has been proposed that this may lead to hysteresis in the Arctic sea ice (e.g., North, 1984), although other feedbacks may also play an important role (e.g., Abbot et al., 2011). To what extent such hysteresis has occurred and/or could occur in the Arctic climate system remains unclear. Some studies suggest a near-linear transition from perennial to seasonal sea ice cover to year-round ice-free conditions (Winton, 2006, 2008). Others suggest a bifurcation in the transition to a seasonally ice-free Arctic (e.g., Abbot et al., 2011; Merryfield et al., 2008). And still others suggest a bifurcation in the transition to a year-round ice-free state (e.g., North, 1984, 1990; Ridley

FIGURE 2.15 Climate Feedback Loop: The melting of Arctic sea ice is an example of a positive feedback loop. As the ice melts, less sunlight is reflected back to space and more is absorbed into the dark ocean, causing further warming and further melting of ice. Source: NRC, 2010b.

et al., 2008; Winton, 2008). Eisenman (2012) discusses how differences in the strength of various climate feedbacks can lead to differences in the likelihood of hysteresis in the Arctic sea ice system.

The numerical evidence for irreversible change to a *year-round* ice-free state was first discussed in studies with simple diffusive climate models (e.g., North, 1984, 1990). In those studies, sea ice exhibits nonlinear behavior such that when it is reduced below a certain threshold (the "Small Ice Cap Instability" threshold), the model sea ice abruptly reverts to year-round ice-free conditions. The change in sea ice is largely irreversible, and substantial cooling is required to reestablish the original sea ice cover. Similar instability is found in numerous models that treat sea ice as a one-dimensional vertical column (Thorndike, 1992; Eisenman and Wettlaufer, 2009; Muller-Stoffels and Wackerbauer, 2011) and some but not all fully coupled IPCC-class climate models (e.g., Winton, 2006; Ridley et al., 2008). Based on these studies, it appears possible, but by no means certain, that a transition to year-round ice-free conditions could result from hysteresis.

Perhaps more relevant for the current Arctic climate is the possibility of a bifurcation to *seasonally* ice-free conditions. Although models generally simulate abrupt summer sea ice loss, the possibility that this might result from a lag in the system is mixed. Indeed, some simulated rapid ice loss events arise from the (random) superposition of large internal variability on the large greenhouse-gas forced trend (Holland et al., 2008). Some single-column model studies reveal bifurcation in the transition to seasonally ice-free conditions (Abbot et al., 2011; Merryfield et al., 2008; Flato and Brown, 1996) but others do not, due to the stabilizing effects of ice thickness on ice growth (Eisenman and Wettlaufer, 2009). A recent study argues that the bifurcation to a seasonally ice-free state in some single column models is an artifact of the model setup (Eisenman, 2012). In general, global climate model studies suggest that a bifurcation to seasonally ice-free conditions is unlikely (Tietsche et al., 2011; Ridley et al., 2008; Winton, 2006; Holland et al., 2008) and that sea ice loss is reversible if greenhouse gas concentrations decline (Armour et al., 2011; Ridley et al., 2012). Note, however, that this regrowth of sea ice would still require a return to the conditions under which sea ice existed; reversing sea ice loss requires reversing Arctic warming. Due to the long-timescales associated with the carbon cycle (e.g., Solomon et al., 2009), reversing sea ice loss would in practice require some type of active carbon removal from the atmosphere.

Regardless of whether hysteresis will occur, the transition to a nearly ice-free Arctic during summer is expected to happen rapidly with rising greenhouse gas forcing. Many methods have been used to predict the timing of near ice-free September conditions, here defined as an Arctic with less than 1 million km^2 of ice extent for the monthly average (see Overland and Wang, 2013 for a review). Extrapolating hindcast model-based historical sea ice volume trends results in September ice-free conditions prior to 2020 (e.g., Maslowski et al., 2012). However, this method does not account for the natural variability in the Arctic system that may be an important factor in trends over the recent past. Studies indicate that this natural variability is sizeable and can lead to a large range of trend magnitudes even on multi-decadal timescales (e.g., Wettstein and Deser, 2013; Kay et al., 2011). Simply extrapolating historical trends also does not account for feedbacks in the system, such as the negative ice thickness-ice growth rate feedback identified by Bitz and Roe (2004) that can slow the ice volume rate of loss. Other methods to predict the timing of September ice-free Arctic conditions have relied on projections from global climate models available through the Coupled Model Intercomparison Project (CMIP3 and CMIP5). A number of techniques have been employed to sub-set or recalibrate these projections based on different aspects of the observed ice cover, including the mean and/or seasonal cycle of ice extent (e.g., Stroeve et al., 2007, 2012a; Wang and Overland, 2009, 2012), historical ice cover trends (Boe et al., 2009), and ice volume and thin ice area (Massonnet et al., 2012). These different methods result in different timings for near-ice free September conditions within the Arctic, which range from the 2030s to 2100. Regardless of when ice-free conditions are actually reached, it is clear that rapid Arctic sea ice loss is already underway and will continue for the foreseeable future.

The impacts of rapid decreases in Arctic sea ice are likely to be considerable (e.g., ACIA, 2005). Ice-free conditions during summer would have potentially large and irreversible effects on various components of the Arctic ecosystem (e.g., Arrigo et al., 2008; Bluhm and Gradinger, 2008; Andersen et al., 2009; Durner et al., 2009). They could have marked impacts on human society and economic development in the coastal polar regions (e.g., Jones et al., 2009; Huntington et al., 2012). Reductions in Arctic sea ice also have implications for Arctic shipping (Smith et al., 2012) and resource extraction (Prowse et al., 2009), which contribute to geopolitical concerns in the region. Additionally, ice-free Arctic summers would lead to large increases in the sensible and latent heat flux into the atmosphere during the fall season that may not only enhance large-scale high latitude terrestrial warming (e.g., Lawrence et al., 2008; Screen and Simmonds, 2010a, b) but also alter the large-scale atmospheric circulation and its variability (Alexander et al., 2004; Seierstad and Bader, 2009; Deser et al., 2010; Screen et al., 2012; Francis and Vavrus, 2012).

In contrast to the Arctic, the Antarctic has seen modest increases in sea ice. Climate models suggest that Antarctic sea ice will decline through the 21st century (e.g., Arzel et al., 2006). But very little work has been done on the stability characteristics of Southern Hemisphere sea ice. Note, however, that while relevant models have shown strong agreement that the warming to date would have reduced Arctic sea ice, there is no such consensus for the Antarctic, with at least some models having simulated sea ice growth in response to moderate warming before switching to sea ice shrinkage from additional warming (e.g., Manabe et al., 1992; see also Liu and Curry, 2010).

Summary and the Way Forward

Arctic sea ice is already changing abruptly with numerous implications for ecosystems, the climate system, and socio-economic impacts. With continued warming, Arctic sea ice will continue to decline. There is considerably less consensus on future changes in Antarctic sea ice. For both poles, understanding future changes requires enhanced monitoring and research efforts.

Monitoring Studies suggest that rapid transitions in the Arctic sea ice are related to the ice thickness distribution (Holland et al., 2006; Lindsay et al., 2009). When the ice becomes thin enough, rapid transitions are likely as large areas of the ice pack can be effectively melted out. As such, monitoring Arctic ice thickness may be useful for predicting rapid changes in sea ice. Satellite-based altimetry measurements, often supplemented by similar measurements from aircraft, hold promise for obtaining basin-scale ice thickness information (e.g., Laxon et al., 2003; Giles et al., 2008; Kwok et al., 2009) and currently provide about a decade-long record. Altimetric freeboard measurements are converted into estimates of total ice thickness assuming isostatic balance. However, there are considerable uncertainties in these estimates due to limited information on the snow conditions on top of the sea ice, the ice density structure, and the high spatial variations in the ice pack. Upward-looking sonar measurements also provide estimates of ice thickness, but have limited spatial and temporal sampling. It remains unknown what uncertainties in ice thickness measurements are "acceptable" to realize any potential predictability for rapid sea ice change.

In addition to sea ice measurements, information on Arctic Ocean conditions may provide insight on the potential for rapid sea ice loss. For example, Holland et al. (2006) related simulated rapid ice loss events to anomalous ocean heat transport into the Arctic from the North Atlantic. Only limited observations of Arctic Ocean conditions currently exist. More research is needed to inform what specific ocean observations in what locations will prove the most useful and potentially enhance our ability to predict rapid ice loss events.

Research It is likely that a rapid transition to seasonally ice-free conditions in the Arctic will occur within the 21st century. However, the repercussions of this for climate, ecosystems, and societal impacts are still uncertain. Additional research is needed in these areas. There is also limited research on the potential predictability of rapid ice loss events. Most studies on ice predictability have used a perfect-model approach, in which a climate model is used to predict conditions simulated by that model, and have focused on seasonal to interannual predictability (e.g., Holland et al., 2011; Blanchard-Wrigglesworth et al., 2011; Chevallier and Salas-Melia, 2012). A recent study has assessed longer-lived ice loss events (Tietsche et al., 2013) and found little predictability in the onset of these events but some predictability in the magnitude of the events in simulations initialized after their onset. However, additional studies using different models and experiment design are needed to determine the robustness of these results.

Provided that rapid losses in sea ice may be predictable, there is additional uncertainty regarding what is required in terms of an observational network and modeling system to predict such events. Observing-network design studies, focused on the issue of abrupt sea ice loss, can be used to inform future observing needs. A recent Arctic Observing Network Design and Implementation Task Force report (AON Design and Implementation Task Force, 2012) provides more details.

A possible transition to year-round ice free conditions is still a distinct possibility. While this would only occur in the more distant future with continued and considerable increases in greenhouse gas concentrations, it would likely have dramatic impacts on the climate and ecosystems. More work is needed to determine why different models exhibit different behavior in this regard.

Finally, very little work has been done on the Antarctic sea ice system in terms of possible abrupt change. This is an additional research need. In particular, a better understanding of mechanisms of Antarctic sea ice variability and change, relevant feedbacks, and ice-ocean-atmosphere interactions is needed. This should be informed by both measurements and modeling of the Antarctic system. Currently, climate models struggle to accurately simulate even the mean conditions of Antarctic sea ice (Turner et al., 2013), and the utility of these models as a tool to study Antarctic ice needs to be critically assessed. It is likely that improvements in models, informed by observations, are needed to better understand the sea ice response to climate forcing and the potential for abrupt change.

ABRUPT CHANGES IN ECOSYSTEMS

Many different biological responses to climate change have been documented, both as an ongoing response to climatic change underway now and in the paleontological record. Thousands of species have reacted to a changing climatic regime by altering their geographic range, abundance, phenology (seasonal patterns), phenotype, or genotype, or in some cases recorded in the fossil record, have become extinct (Barnosky, 1986, 2009; Barnosky et al., 2003; Blois and Hadly, 2009; Brook and Barnosky, 2012; Hadly et al., 2004; Harnik et al., 2012; Pandolfi et al., 2011; Parmesan, 2006; Parmesan and Yohe, 2003; Root et al., 2003; Walther et al., 2002; Moritz et al., 2008). The ubiquity of biological response to climate indicates that climate changes underway will cause existing ecosystems to change noticeably. There is a possibility that at the ecosystem level the climatically-triggered changes will include abrupt state changes within the next few decades. This is supported by a large body of empirical and theoretical work that demonstrates when ecosystems change states, whatever the ultimate driver of change, they tend to do so abruptly once particular thresholds are crossed (for instance, see Barnosky et al., 2012; Bascompte and Sole, 1996; Carpenter et al., 2011; Peters et al., 2009; Scheffer et al., 2009; Swift and Hannon, 2010; and references therein).

Such abrupt state changes are well-documented for ecosystems at many scales, and can be triggered by a variety of forcing factors—including pollution, resource extraction, deforestation, and other land use changes—with climate change being only one of them (Scheffer et al., 2009; Lenton et al., 2008; Barnosky et al., 2012). In some cases ecosystems are known to have changed from the "old" to the "new" state within decades. For instance, in southern New Mexico, a changing dynamic between wind, water, and animals caused grasslands to transform into less productive shrublands over a total of about 70 years, with the shift from predominantly grassland to predominantly shrubland bracketed between 1980 and 1990 (Peters et al., 2009). More broadly, grazing and fire suppression have contributed to historic transitions from semi-arid grassland to desert shrub vegetation regimes in many parts of the southwestern United States and Africa (e.g., Schlesinger et al., 1990; Holdo et al., 2009), and prescribed fires as a land management practice can lead to abrupt transition from a savanna-desert plant regime to a savanna-grassland regime (e.g., Taylor et al., 2012).

Climate change has been shown to be an important component of abrupt ecosystem state-changes. A particularly instructive example is the Sahel region of Africa, which switched from vegetated land that supported cattle to unproductive desert within 5 years beginning about 1965, causing widespread famine and an international crisis in the region that continues to be a problem today. In the Sahel the onset of desertification involved an interplay between the position of the West African Monsoon

(Lenton et al., 2008) and the local climate feedbacks that are controlled by the amount of vegetative cover (Stewart, 2010). In the case of the 1960s desertification, a few bad drought years were caused by warm sea surface temperatures, which weakened the influence of the South African monsoon over the Sahel. At the same time, grazing pressure denuded the landscape. This, in turn, regionally increased albedo (caused more sunlight to be reflected from the land) and reduced evaporation, which further weakened the monsoon (Stewart, 2010). As a result, some regions of the Sahel that formerly supported grazing still remain unproductive desert today. Other empirical data and theoretical models indicate that such interplays between land use and climate change are likely to cause desertification in many other dryland environments that currently support 2 billion people (D'Odorico et al., 2013; Schlesinger et al., 1990).

While much of the work on whole-ecosystem regime shifts has focused on drylands, a growing body of evidence indicates abrupt state shifts will plausibly affect many other ecosystems as climate continues to change over the next several decades. Boreal forests appear susceptible to rapid transition to sparse woodland or treeless landscapes as temperature and precipitation patterns shift (Scheffer et al., 2012b). Climatic shifts would be expected to exacerbate the large-scale ecosystem changes in boreal regions that human induced changes from grazing or fires can also trigger (e.g., Chapin et al., 2004, Randerson et al., 2006).

At the global scale, observations show that the transitions from forests to savanna and from savanna to grassland tend to be abrupt when annual rainfall ranges from 1,000 to 2,500 mm and from 750 to 1,500 mm, respectively (Hirota et al., 2011; Mayer and Khalyani, 2011; Staver et al., 2011). Such rainfall regimes cover nearly half of the global land, where either a gradual climate change across the ecosystem thresholds or a strong perturbation due to either extreme climate events, land use, or diseases could trigger abrupt ecosystem changes. The latter could in turn amplify the original climate change in the areas where land surface feedback is important to climate (e.g., Friedlingstein et al., 2006; Scheffer et al., 2006).

Amazon forests represent the world's largest terrestrial biome and potentially the tropical ecosystem most vulnerable to abrupt change in response to future climate change in concert with agricultural development (e.g., Cox et al., 2000; Lenton et al., 2008; Zelazowski et al., 2011). Thus, the rest of this section explores the risk of collapse of the Amazon forests as an example of a potentially vulnerable ecosystem.

Abrupt Transformation of the Amazon Forest by Climate Change and Deforestation

The closed-canopy equatorial forests of Amazonia are iconic in public perception: lush, highly productive, richly diverse ecosystems. The forests are characterized by a tall canopy of broadleaved trees, 30-40m high, sometimes with impressive emergent trees up to 55 m or taller. The Brazilian portion of the Amazon comprises 4×10^6 km^2,[12] less than 1 percent of global land area, but disproportionally important in terms of above-ground terrestrial biomass (15 percent of global terrestrial photosynthesis [Field et al., 1998]) and number of species (~25 percent, Dirzo and Raven, 2003). Direct human intervention via deforestation represents an existential threat to this forest: despite recent moderation of rates of deforestation, the Amazon forest is on track to be 50 percent deforested within 30 years—arguably by itself an abrupt change of global importance (Fearnside, 1983; Gloor et al., 2012).

Climate change represents yet another source of stress on an already distressed system. In particular, seasonal and multi-year drought frequency and intensity may have increased, and such increase could in part be attributable to anthropogenic forcing (e.g., Dai, 2011; Li et al., 2008). The projected radiative-forced increase of extreme surface temperatures and stronger spring barrier for wet season onset (Cook et al., 2010a; Seth et al., 2011) would increase risk of forest fires (Golding and Betts, 2008), although how changes of ENSO, AMO, and aerosols loadings will influence future droughts remain unclear (e.g., Andreae et al., 2005).

Biophysical Mechanisms and Feedbacks Defining the Boundaries of the Closed-Canopy Equatorial Forest At the continental scale, nonlinear feedbacks between the equatorial forest and the atmosphere have been recognized for decades. These forests receive enormous inputs of radiant energy and moisture. A significant fraction (25-35 percent) of regional rainfall represents water recycled between the forest and the atmosphere (Salati et al., 1979; Eltahir and Bras, 1994; Zeng et al., 1996; da Rocha et al., 2009), providing a strong homeostatic mechanism, i.e., the forest can regenerate rainfall that waters itself. The local water recycling provides nearly 100 percent of the regional rainfall during dry season (Li and Fu, 2004). Thus, forest clearing would reduce dry season rainfall, increase fire risk, and possibly delay wet season onset (Gash and Nobre, 1997; Fu and Li, 2004, Costa and Pires, 2010). This, in turn, increases the ecosystem's vulnerability to forest clearing via reduction in water recycling as forest cover is removed (Salati and Nobre, 1991), or via changes in rain formation processes (e.g.,

[12] Approximately 40% of the total area of the United States.

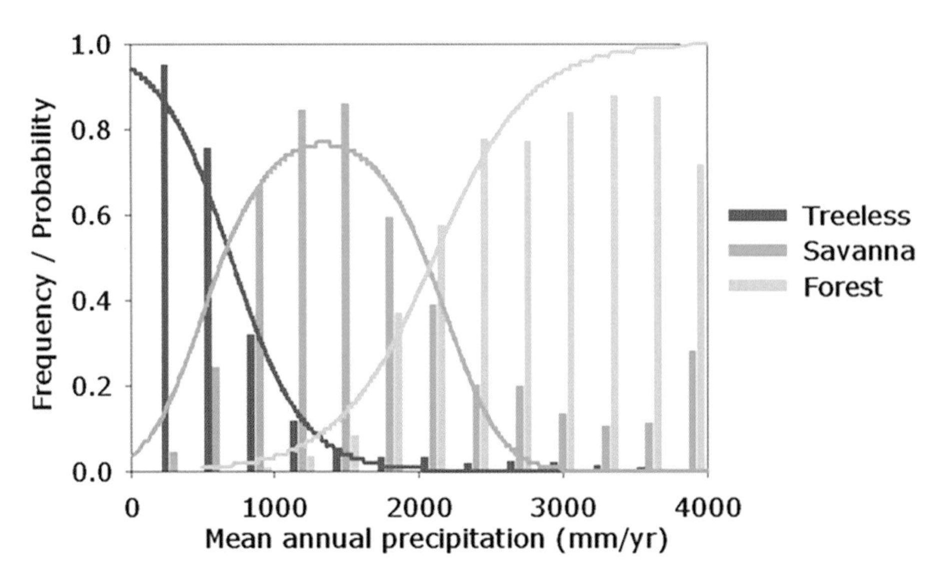

FIGURE 2.16 Frequency of tropical forests and savannas, plotted vs. mean annual precipitation (Hirota et al., 2011).

Andreae et al., 2004) and monsoon circulation transition (Zhang et al., 2009, Bevan et al., 2009) due to inputs of smoke from agricultural burning.

Nonlinear feedbacks, thresholds, and bi-stable hysteresis[13] operate at the ecosystem scale. Classic biogeographical studies have shown that shifts in the balance between potential evaporation and precipitation ("potential evaporation ratio") can give rise to threshold behavior in these forests (Holdridge, 1947, 1964). With current mean temperatures in the range 26-29°C in Amazonia, the lower limit of rainfall to sustain a closed canopy forest is about 1,600 mm/yr (Nix, 1983; Hirota et al., 2011; see Figure 2.16). The length of the dry season and rainfall variability are also important, however. Tropical or subtropical areas with higher rainfall, even with seasonally flooded landscapes, typically have fire-adapted, low-stature ecosystems if the dry season is longer than ~5 months (e.g., Pantanal in Brazil, northern Australia).

Biophysical Mechanisms and Feedbacks of Nonlinear Feedbacks, Thresholds, and Hysteresis Trees that make up the main canopy, or emerge above it, access moisture

[13] A bi-stable hysteresis refers to a system that can be in one of two stable states and which state it occupies depends on the history of the system. An example would be a landscape that has been stable in the past as either a rain forest or a dry savanna depending on the pervading climate conditions.

stored deep in the soil and use it to harvest the intense sunlight available in the dry season. At least in eastern Amazonia, where deep soil columns are accessible to roots (see, for example, Zeng et al., 1996; Kleidon and Heimann, 2000; Nobre and Borma, 2009), transpiration rates in tropical forests are sustained at peak levels throughout dry periods of four to five months. This persistence of very high latent heat flux is opposite to the seasonal trends of transpiration and evaporation in savannas, even with similar total rainfall (Saleska et al., 2003; da Rocha et al., 2009). The forest is resilient, with rapid growth of canopy trees. Proliferation of diverse flora and fauna is enabled by the moderated microclimate, made possible by conversion of solar energy to latent heat.

Two processes can disrupt this system and cause a transition to a less dense transitional forest or to fire-adapted vegetation: (1) damage to canopy trees by fires fueled by dry understory vegetation; or (2) depletion of deep stores of soil moisture. These processes can potentially lead to *abrupt* changes in Amazon forest structure and extent.

1. Fires—Natural fires are rare in closed-canopy forests, but human-set fires are common today, and may have been widespread in the past when indigenous populations were much higher than today (Roosevelt et al., 1996; Bush et al., 2000; Shepard et al., 2012). Fires that enter the closed-canopy forest from agricultural or forest clearing activities are normally of low intensity, but may kill large trees during droughts or strong dry seasons. Smoldering ground fires may also be lethal if the fires return too frequently (Cochrane et al., 1999; Numata et al., 2010). Hence the damage to canopy trees by fires depends on the factors that control intensities and return intervals: temperature, humidity, duration of the dry season, occurrence of dry season rainfall, sources of ignition, and fuel loads (Adams, 2013). Positive feedbacks operate for some time after ground fire, because the killing of undergrowth vegetation creates more dry fuel, higher temperatures, and lower humidity for the next fire (Cochrane et al., 1999). Pueyo et al. (2010) found that strong positive feedbacks between drought and deforestation caused the very large fires in Roraima, Brazil, in 1997.

Although apparently irreversible shifts in ecosystem composition may occur after fires, the underlying causes are often changes in dry season balances of energy and water (means), the frequency of extremely hot or dry periods (variances), edge effects at forest-agriculture boundaries, and the occurrence of sources of ignition (human associated, e.g., agriculture). Vulnerable forest may persist for extended periods until an "event" leads to an actual transition, which may appear to be irreversible.

2. Deep water depletion and variability of rainfall in vulnerable (near-threshold) forests—Nepstad et al. (2007) carried out a manipulation experiment on a central Amazon forest by reducing rainfall in the wet season by 60 percent. Initially the forest was unaffected, but after the third year the largest trees began to die. Detailed simulations support Nepstad's hypothesis that the key factor was incomplete recharge of deep soil moisture, which supports prodigious rates of photosynthesis and transpiration by the largest trees during the dry season (Figure 2.17; Ivanov et al., 2012). Hutyra et al. (2005) used the 100-year, monthly reconstruction of rainfall by New et al. (1999) to refine the Holdridge plot (Holdridge, 1947, 1964) and explore the role of

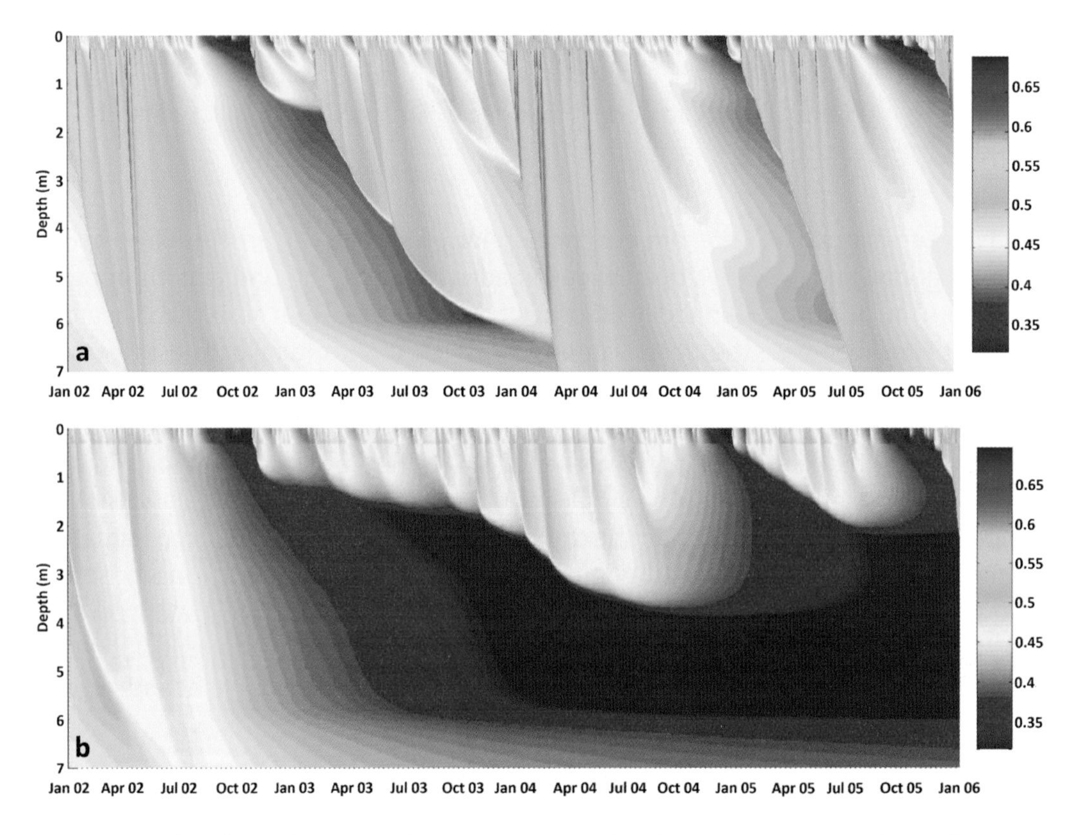

FIGURE 2.17 Colors show volumetric soil moisture content, where blue is more moist and red is drier. Soil moisture shown as a function of depth for two model simulations over a four-year period. (*a*). Soil moisture profile in a central Amazon forest, simulated using observed meteorology. Trees in the upper canopy had a higher fraction of deep roots, conversely for understory plants. (*b*) Simulation with 60 percent reduction in precipitation or increase in evaporation; upper canopy trees cannot survive. Source: Ivanov et al., 2012 (model) using data from Nepstad et al., 2007.

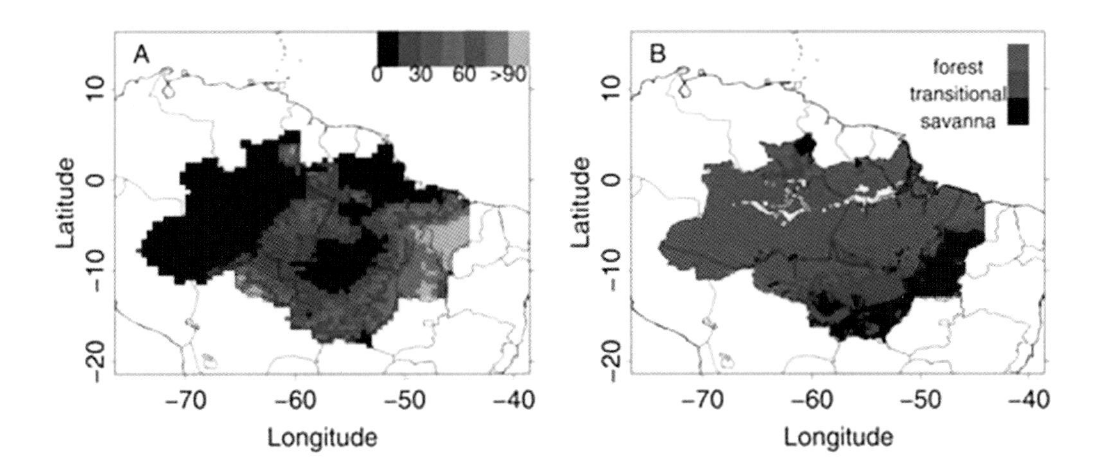

FIGURE 2.18 (*left*) Drought probability (soil moisture < 25 percent of capacity) for 100 years of CRU monthly climate reconstruction, based on evaporation parameterized from eddy flux data and plant available water capacity from biogeographical information. The boundary of the closed canopy forest with transitional forest or savanna (*right*) corresponds closely to 50 percent drought probability, consistent with the requirement to recharge deep soil reservoirs every 2 years as discussed by Nepstad et al. (2007). Source: Hutyra et al., 2005.

variance in determining the stability of tropical forests, finding that the natural forest-savanna boundary corresponded to a 50 percent annual probability of incomplete recharge (Figure 2.18). Thus observed biogeographical boundaries give independent support for Nepstad's hypothesis.

It appears that forests in the Amazon, at least in the central and eastern regions, may be rendered vulnerable to collapse either by increases of Potential Evaporation (PE, by increasing temperature or sunlight) or decreasing precipitation (Pc). In areas close to the biogeographic boundary, increasing variability of rainfall or longer dry seasons can shift forests to savannas, without changes in mean PE or Pc. In general, vulnerable forests may persist for extended periods until events, such as a series of strong droughts or repeated fire occurrences, lead to ecosystem collapse.

It is unclear how much hysteresis would attach to the re-establishment of tall trees in a closed canopy forest that had collapsed due to climatic shifts, since little is known about how this process proceeds. Amazonian forests appear to have expanded during moist periods of the Holocene and contracted in dry periods (Oliveira and Marquis, 2002), but in at least one case, the forest did not re-occupy its previous extent for some time after wetter conditions returned (Ledru et al., 1998). This "hysteresis" could have been associated with any or all of a number of mechanisms, including persistent flammability, inhibited recruitment of big trees in locations where their early growth was not sheltered so that their roots could reach deep soil water, persistent shrub vegetation types, etc.

Summary and the Way Forward

Lenton et al. (2008) and Nobre and Borma (2009) have summarized current understanding of "tipping points" in Amazonian forests. Global and regional models do indeed simulate hysteresis and collapse of Amazonia forests. Models exhibit these shifts for a range of perturbations: temperature increases of 2-4°C, precipitation decreases by ~40 percent (1100 mm, according to Lenton et al., 2008), and/or deforestation that replaces large swathes of the forest with agriculture (e.g., Betts et al., 2004).

It is noted, however, that large-scale models do not have the detailed representation of subsurface hydrological processes that the Nepstad et al. (2007) data would seem require (see Ivanov et al., 2012). Thresholds may occur much closer to current conditions, for example, if precipitation falls below 1,600-1,700 mm (Nobre and Borma, 2009). Indeed, long-lasting damage to Amazonian forests may have occurred after the single severe drought in 2005 (Saatchi et al., 2013). In addition, the large-scale climate models have large errors in representing the processes that control rainfall variability and changes (e.g., Li et al., 2006; Yin et al., 2012). This is also a leading source of uncertainty in determining the risk of the Amazon ecosystem collapse (Friedlingstein et al., 2006; Good et al., 2013). A recent study shows that the dry season length over part of the Amazonia has increased much faster than that represented by climate models for both the current and future climate (Fu et al. 2013). Thus, the risk of rainforest collapse due to climatic drying is likely significantly higher than that represented by the current climate models. The committee concludes that credible possibilities of thresholds, hysteresis, indirect effects, and interactions amplifying deforestation, make abrupt (50 year) change plausible in this globally important system. Rather modest shifts in climate and/or land cover may be sufficient to initiate significant migration of the ecotone defining the limit of equatorial closed-canopy forests in Amazonia, potentially affecting large areas.

Monitoring for early warning The very strong 2005 drought gave rise to intensive studies of forest resilience and vulnerability using multi-spectral imaging from the MODIS satellite instrument (Marengo et al., 2008; Saleska et al., 2007), from the QuickScat active microwave sensor (Saatchi et al., 2013), and from distributed forest plots

(Phillips et al., 2009). The forest initially "greened up" during the dry period, as anticipated from seasonal changes under normal conditions (da Rocha et al., 2009), but subsequently experienced productivity decline and mortality (Phillips et al., 2009). Notably, the microwave data indicated that the forest did not fully recover for at least three years (Saatchi et al., 2013). The combined all-weather canopy surface temperature provided by passive microwave sensor, such as AMSR-E and hyperspectral data are also important for monitoring plants water stress for early warning.

Hence it appears that the tools required for monitoring and provision of early warning are at hand. Multi-spectral and active microwave data from satellites, plus an effective network of ecological plots, appear capable of monitoring response to climate change. Landsat combined with multi-spectral satellite sensing and LIDAR (e.g., Asner et al., 2010, 2012), can detect forest clearing and chart regrowth. It is not clear, however, that these tools will actually be available in the future to provide the required data at high spatial and temporal resolution with the necessary continuity. Commitment to carrying forward the satellite sensors is in doubt.

Long-term networks of ecological plots should be the foundation of global change studies of forest ecosystems. Current networks in the tropics (e.g., the Rainfor network[14]), represent diverse collections of scientists from many countries pursuing a range of questions, funded by a patchwork of sources. There is a very strong need for a "global service" network that makes comprehensive monitoring and early detection its main focus in the near future. This outcome would require a framework that does not now exist: a well-structured organization with long-term funding, broad international participation, and quality controlled data that enter the public domain. The lack of such an effort today undermines efforts to detect and respond to ecological changes in tropical systems, both forest and non-forest.

Extinctions: Marine and Terrestrial

Extinction is an irreversible biological change that can fundamentally alter the ecosystem of which a lost species was a part, contributing to ecological state shifts as described in the last section and to depleting ecosystem services as described below (see Chapter 3, Boxes 3.1 and 3.2). In the context of this report, extinction is recognized as "abrupt" in two respects. First, the numbers of individuals and populations that ultimately compose a species may fall below critical thresholds such that the likelihood for species survival becomes very low. This kind of abrupt change is often cryptic, in

[14] http://www.rainfor.org.

that the species at face value remains alive for some time after the extinction threshold is crossed, but becomes in effect a "dead clade walking" (Jablonski, 2001). Such losses of individuals that take species towards critical viability thresholds can be very fast—within three decades or less, as already evidenced by many species now considered at risk of extinction due to causes other than climate change by the International Union for the Conservation of Nature.[15] The second kind of abrupt change is simply the terminal event in the extinction process—the loss of the last individual of a species. While this is what most people recognize as extinction, it generally postdates by decades the dropping of numbers of individuals below species-viability thresholds. The abrupt impact of climate change on causing extinctions of key concern, therefore, is its potential to deplete population sizes below viable thresholds within just the next few decades, whether or not the last individual of a species actually dies.

The possibility that ongoing anthropogenic climate change will push many species past extinction thresholds is increasingly cited (Barnosky et al., 2011; Foden et al., 2013; Harnik et al., 2012; NRC, 2011a; Pimm, 2009; Cahill et al., 2012) and is rooted in considerations about both rate and amount of projected change. The rate of global climate change now underway is at least an order of magnitude faster than any warming event in the last 65 million years (Barnosky et al., 2003; Blois and Hadly, 2009; Blois et al., 2013; Diffenbaugh and Field, 2013).[16] From the late 20th to the end of the 21st century, climate has been and is expected to continue changing faster than many living species, including humans and most other vertebrate animals, have experienced since they originated. Consequently, the predicted "velocity" of climate change—that is, how fast populations of a species would have to shift in geographic space in order to keep pace with the shift of the organisms' current local climate envelope across the Earth's surface—is also unprecedented (Diffenbaugh and Field, 2013; Loarie et al.,

[15] http://www.iucnredlist.org/.

[16] In this context we refer to global-scale warming rates, not regional or local. At the global scale, glacial-interglacial transitions exhibit the most rapid and highest-magnitude warming rates documented in Earth history (Barnosky et al., 2003; Blois and Hadly, 2009; Diffenbaugh and Field, 2013). Note that regional proxies, such as the oxygen-isotope temperature reconstructions from the Greenland Ice Core Project that record Dansgaard-Oeschger events, often indicate faster regional rates of climate change than the overall global average for glacial-interglacial transitions, just as today warming is more pronounced in Arctic regions than in equatorial regions (Barnosky et al., 2003; Diffenbaugh and Field, 2013). Therefore in comparing rates of global warming today with past rates of global warming, it is essential to use global averages, rather than comparing a global average with a regional proxy. In so doing, Diffenbaugh and Field (2013) estimate that climate change now is proceeding at "at a rate that is at least an order of magnitude and potentially several orders of magnitude more rapid than the changes to which terrestrial ecosystems have been exposed during the past 65 million years." Blois and Hadly (2009) and Barnosky et al. (2003) discuss how even when standardizing for higher rates that are apt to be a statistical artifact of measuring rates over shorter time intervals, the current global average rate of warming is above the global average that species have experienced over the past 65 million years.

2009). Moreover, the overall temperature of the planet is rapidly rising to levels higher than most living species have experienced (Figure 2.19). Consequently all the populations in some species, and many populations in others, will be exposed to local climatic conditions they have never experienced (so-called "novel climates"), or will see the climatic conditions that have been an integral part of their local habitats disappear ("disappearing climates") (Williams et al., 2007). Models suggest that by the year 2100, novel and disappearing climates will affect up to a third and a half of Earth's land surface, respectively (Williams et al., 2007), as well as a large percentage of the oceans (see, for example, the Ocean Acidification section of this report; NRC, 2011a; Ricke et al., 2013). Thus, many species will experience unprecedented climatic conditions across their geographic range. If those conditions exceed the tolerances of local populations, and those populations cannot migrate or evolve fast enough to keep up with climate change, extinction will be likely. These impacts of rapid climate change will moreover occur within the context of an ongoing major extinction event that has up to now been driven primarily by anthropogenic habitat destruction.

Most projections of future climate-driven extinctions rest upon the assumption that potential geographic distribution of each species is ultimately determined by the climatic tolerances of the populations that make up that species. These tolerances define a species "climate envelope" which moves in space as the global climate changes,

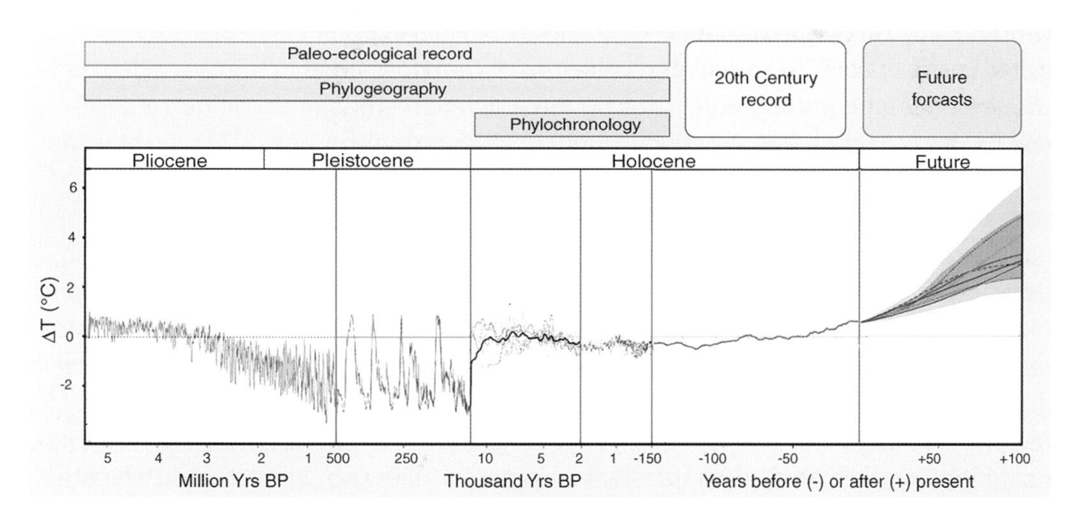

FIGURE 2.19 Global climatic conditions (here exemplified by temperature rise) by 2050-2100 are expected to be outside the range that most living species have ever experienced (figure from Moritz and Agudo, 2013).

causing the decline of populations at the trailing edge. If a species' populations cannot adapt fast enough to tolerate local climate change, or migrate fast enough to track the changing geographic location of suitable climate space at the leading edge of the species range, that species will go extinct (Aitken et al., 2008; Corlett and Westcott, 2013). Species distributions have shifted across the landscape in response to past climate change without evidence of climate-driven elevated extinction rates (Moritz and Agudo, 2013; Jackson and Weng, 1999; Sandel et al., 2011). However, those past climate changes were considerably slower and less intense than what species are expected to experience over the next 30 to 80 years, projections which lead to forecasts of significant future extinctions (Moritz and Agudo, 2013). For example, recent work suggests that up to 41 percent of bird species, 66 percent of amphibian species, and between 61 percent and 100 percent of corals that are not now considered threatened with extinction will become threatened due to climate change sometime between now and 2100 (Foden et al., 2013; Ricke et al., 2013), and that in Africa, 10-40 percent of mammal species now considered not to be at risk of extinction will move into the critically endangered or extinct categories by 2080, possibly as early as 2050 (Thuiller et al., 2006).

An important consideration for such projections is the spatial velocity of climate change (Diffenbaugh and Field, 2013; Loarie et al., 2009). At the last glacial-interglacial transition (the most rapid global climatic transition known prior to today's), the range of plant dispersal velocities was between 0.1 and 1.0 km/yr (Loarie et al., 2009), with some species lagging behind their moving climate envelope (Normand et al., 2011; Ordonez, 2013). Predicted climate velocities for the next century are considerably faster. Loarie et al. (2009) calculated velocities of climate change in terms of relative changes in temperature gradients using three different emissions scenarios (A2, AB, and B1) and concluded that between 2050 to 2100, organisms now living in areas that cover about 29 percent of the planet's land will have to disperse faster than observed post-glacial velocities. Diffenbaugh and Field (2013), using the more recent RCP8.5 scenario, expressed velocities in terms of nearest equivalent temperatures. Their calculations indicated that by 2081-2100, most terrestrial species on the globe would need to disperse at a rate that exceeds 4 km/yr, and that nearly half of the land surface would require dispersal velocities that exceed 8 km/yr. Over roughly a third of Earth's lands, dispersal velocities would need to exceed 16 km/yr, with velocities in high-latitude regions reaching more than 128 km/yr (Figure 2.20). All of these estimates assume no inhibition of dispersal, with transient trajectories following unobstructed "climate paths" to the predicted future climate space. However, short-term climate fluctuations may result in transient loss of suitable climate for certain species, thus preventing those species from migrating to track suitable future conditions (Early and Sax, 2011).

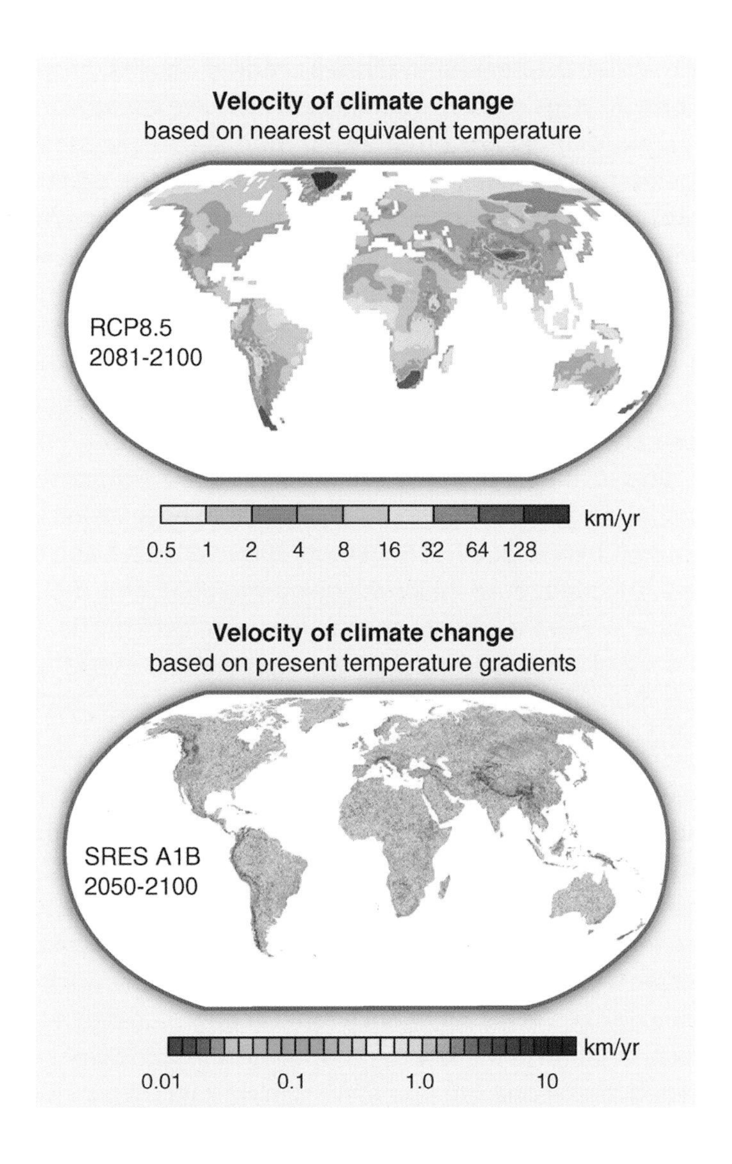

FIGURE 2.20 As temperatures rise, populations of many species will have to move to new habitats to find suitable food, water, and shelter. The colors on these maps show how fast individuals in a species will have to move across the landscape in order to track the mean temperature that now characterizes the places where they live. The figure shows two methods of calculating the velocity of climate change for different time periods at the end of this century. The top panel shows the velocity in terms of nearest equivalent temperature, i.e., the climate change velocity in the CMIP5 RCP8.5 ensemble, calculated by identifying the closest location (to each grid point) with a future annual temperature that is similar to the baseline annual temperature. The lower panel expresses velocity as change in present temperature gradients calculated by using the present temperature gradient at each location and the trend in temperature projected by the CMIP3 ensemble in the SRES A1B scenario. Source: Diffenbaugh and Field, 2013.

Will species be able to track future rapid climate change? Species range shifts in response to the past 50 years or so of warming climate have already been observed (Chen et al., 2011; Parmesan, 2006; Parmesan and Yohe, 2003; Poloczanska et al., 2013; Root et al., 2003). Recent meta-analyses indicate that on average, examined terrestrial species have been moving poleward about 1.76 km/yr (reported as 17.6 $\pm$ 2.9 km/decade), apparently keeping pace with regional temperature change, although species range shifts to higher elevations have on average lagged behind climate (Chen et al., 2011). However, individual species vary widely in observed dispersal velocity (Chen et al., 2011), and several studies report many plant populations lagging behind recent warming (Zhu et al., 2012, Corlett and Westcott, 2013). Marine species have been moving poleward at about 7.2 km/yr (reported as 72 $\pm$ 13.5 km/decade) (Poloczanska et al., 2013). The faster rates in marine organisms may occur because dispersal is enhanced by ocean currents. It is unknown whether the species that have been exhibiting a range-shift response (Chen et al., 2011; Parmesan, 2006; Parmesan and Yohe, 2003; Poloczanska et al., 2013; Root et al., 2003) will be able to accelerate their dispersal velocities to keep pace with the climate change expected over the next few decades under business-as-usual scenarios. This is an area that requires more investigation. We also need to know more about the role of evolutionary adaptation in shaping future species range shifts (Hoffmann and Sgro, 2011).

It is an open question whether the climatic tolerances of local populations can evolve fast enough to keep up with rapid climate change (Aitken et al., 2008; Hoffmann and Sgro, 2011; Moritz and Agudo, 2013). Rapid phenotypic evolution may be required to track changing conditions. For example, adaptation of Sitka spruce to climate change projected for 2080 under the A2 scenario (using the Canadian and Hadley GCMs) would require advancement of annual bud set date within each local population by over 50 days (Aitken et al., 2008). Such rapid change can only occur if there is sufficient genetic variation in the selected population. In some cases, extensive gene flow or assisted migration from populations in warmer parts of the range may enhance the genetic potential for rapid evolutionary response of poleward populations (e.g., Kuparinen et al., 2010). However, adaptive evolution of populations at the warmer range limit of a species will be limited by the amount of genetic variation and covariation within populations for traits affecting climate tolerance (Hoffmann and Sgro, 2011; Shaw and Etterson, 2012). In general, "evolutionary rescue" from extinction pressures induced by climate change requires large populations and high levels of genetic variation for natural selection to act upon (Alberto et al., 2013; Moritz and Agudo, 2013), and may thus be unlikely to occur in many species, particularly rare endemics, and species whose genetic variation has already been severely decimated by other extinction pressures (for instance, species like tigers and black rhinos). If adaptive evolution lags behind the rate of environmental change, population viability will decline, increasing the risk of local and global extinction (Aitken et al., 2008; Maurer, 1999; Stephens et al., 1999).

A critical consideration is that the biotic pressures induced by climate change will interact with other well-known anthropogenic drivers of extinction to amplify what are already elevated extinction rates. Even without putting climate change into the mix, recent extinction has proceeded at least 3-80 times above long-term background rates (Barnosky et al., 2011) and possibly much more (Pimm and Brooks, 1997; Pimm et al., 1995; WRI, 2005),[17] primarily from human-caused habitat destruction and overexploitation of species. The minimally estimated current extinction rate (3 times above background rate), if unchecked, would in as little as three centuries result in a mass extinction equivalent in magnitude to the one that wiped out the dinosaurs (Barnosky et al., 2011) (see Box 2.4). Importantly, this baseline estimate *assumes no effect from climate change*. A key concern is whether the added pressure of climate change would substantially increase overall extinction rates such that a major extinction episode would become a *fait accompli* within the next few decades, rather than something that potentially would play out over centuries.

Known mechanisms by which climate change can cause extinction include the following.

1. Direct impact of an abrupt climatic event—for example, flooding of a coastal ecosystem by storm surges as by seas rise to levels discussed earlier in this report.
2. Gradually changing a climatic parameter until some biological threshold is exceeded for most individuals and populations of a species across its geographic range—for example, increasing ambient temperature past the limit at which an animal can dissipate metabolic heat, as is happening with pikas at higher elevations in several mountain ranges (Grayson, 2005). Populations of ocean corals (Hoegh-Guldberg, 1999; Mumby et al., 2007; Pandolfi et al., 2011; Ricke et al., 2013) and tropical forest ectotherms (Huey et al., 2012) also inhabit environments close to their physiological thermal limits and may thus be vulnerable to climate warming. Another potential threshold phenomenon is decreasing ocean pH to the point that the developmental pathways of many invertebrates (NRC, 2011a; Ricke et al., 2013) and vertebrate species are disrupted, as is already beginning to happen (see examples below).

[17] The wide range of estimates for exactly how much extinction rates are now elevated is because there is much statistical uncertainty in estimating the background rate from fossils, even in the best cases where the fossil record is reasonably good (as for mammals).

3. Interaction of pressures induced directly by climate change with non-climatic anthropogenic factors, such as habitat fragmentation, overharvesting, or eutrophication, that magnify the extinction risk for a given species—for example, the checkerspot butterfly subspecies *Euphydryas editha bayensis* became extinct in the San Francisco Bay area as housing developments destroyed most of their habitat, followed by a few years of locally unfavorable climate conditions in their last refuge at Jasper Ridge, California (McLaughlin et al., 2002).
4. Climate-induced change in biotic interactions, such as loss of mutualist partner species, increases in disease or pest incidence, phenological mismatches, or trophic cascades through food webs after decline of a keystone species. Such effects can be intertwined with the intersection of extinction pressures noted in mechanism 3 above. In fact, the disappearance of checkerspot butterflies from Jasper Ridge was because unusual precipitation events altered the timing of overlap of the butterfly larvae and their host plants (McLaughlin et al., 2002).

BOX 2.4 MASS EXTINCTIONS

Mass extinctions are generally defined as times when more than 75 percent of the known species of animals with fossilizable hard parts (shells, scales, bones, teeth, and so on) become extinct in a geologically short period of time (Barnosky et al., 2011; Harnik et al., 2012; Raup and Sepkoski, 1982). Several authors suggest that the extinction crisis is already so severe, even without climate change included as a driver, that a mass extinction of species is plausible within decades to centuries. This possible extinction event is commonly called the "Sixth Mass Extinction," because biodiversity crashes of similar magnitude have happened previously only five times in the 550 million years that multi- cellular life has been abundant on Earth: near the end of the Ordovician (~443 million years ago), Devonian (~359 million years ago), Permian (251 million years ago), Triassic (~200 million years ago), and Cretaceous (~66 million years ago) Periods. Only one of the past "Big Five" mass extinctions (the dinosaur extinction event at the end of the Cretaceous) is thought to have occurred as rapidly as would be the case if currently observed extinctions rates were to continue at their present high rate (Alvarez et al., 1980; Barnosky et al., 2011; Robertson et al., 2004; Schulte et al., 2010), but the minimal span of time over which past mass extinctions actually took place is impossible to determine, because geological dating typically has error bars of tens of thousands to hundreds of thousands of years. After each mass extinction, it took hundreds of thousands to millions of years for biodiversity to build back up to pre-crash levels.

These dangers of extinction from climate change are well documented for mammals, birds, reptiles, amphibians (Foden et al., 2013; Pimm, 2009; Sinervo et al., 2010), and corals (Hoegh-Guldberg, 1999; Mumby et al., 2007; Pandolfi et al., 2011; Ricke et al., 2013). Theoretical considerations and some empirical data also indicate that continued climate change at its present pace would be detrimental to many species of marine clams and snails, fish, tropical ectotherms, and some species of plants (examples and citations below). For such species, continuing the present trajectory of climate change would very likely result in extinction of most, if not all, of their populations by the end of the 21st century. The likelihood of extinction from climate change is low for species that have short generation times, produce prodigious numbers of offspring, and have very large geographic ranges. However, even for such species, the interaction of climate change with habitat fragmentation may cause the extirpation of many populations. Even local extinctions of keystone species may have major ecological and economic impacts.

The interaction of climate change with habitat fragmentation has high potential for causing extinctions of many populations and species within decades (before the year 2100 if not sooner). The paleontological record and historical observations of species indicate that in the past species have survived climate change by their constituent populations moving to a climatically suitable area, or, if they cannot move, by evolving adaptations to the new climate. The present condition of habitat fragmentation limits both responses under today's shifting climatic regime. More than 43 percent of Earth's currently ice-free lands have been changed into farms, rangelands, cities, factories, and roads (Barnosky et al., 2012; Foley et al., 2011; Vitousek et al., 1986, 1997), and in the oceans many continental-shelf areas have been transformed by bottom trawling (Halpern et al., 2008; Jackson, 2008; Hoekstra et al., 2010). This extent of habitat destruction and fragmentation means that even if individuals of a species can move fast enough to cope with ongoing climate change, they will have difficulty dispersing into suitable areas because adequate dispersal corridors no longer exist. If individuals are confined to climatically unsuitable areas, the likelihood of population decline is enhanced, resulting in high likelihood of extinction if population size falls below critical values, from processes such as random fluctuations in population size (Maurer, 1999) or Allee effects (Stephens et al., 1999).

These considerations make it very likely that at least some populations and species would likely go extinct, and even more will likely drop below viable numbers of individuals within the next few decades simply because they could not disperse across fragmented landscapes fast enough to keep pace with movement of their required climate zones. Concerted efforts of human-mediated translocation of species could help mitigate this, but the practice is still regarded as controversial and experimental

(McLachlan et al., 2007; Ricciardi and Simberloff, 2009; Hoegh-Guldberg et al., 2008; Sax et al., 2009; Schwartz et al., 2012).

Vulnerabilities of Species to Extinction

The demonstrable vulnerabilities of populations and species to extinction by climate change fall into three categories.

1. Those whose physiological tolerances to various climatic parameters will be exceeded by climate change throughout their geographic range.
2. Those that will have their growth, development, reproduction, or survival detrimentally impacted by climate change or consequent changes in biotic interactions, resulting in population decline.
3. Those that are effectively trapped by habitat fragmentation in areas where climate changes detrimentally, even though suitable climatic habitat may exist for them elsewhere in the world.

Examples of species in Category 1 are: polar bears, which require sea ice in order to thrive, as their primary hunting strategy to maintain adequate fat reserves is waiting for seals to emerge from openings in the ice (Derocher et al., 2004); mountain species such as pikas (Grayson, 2005; Beever et al., 2011), which cannot survive sustained temperatures above ~27°C (80°F); endemic Hawaiian silverswords, which are restricted to cool temperatures at high altitudes and die from moisture stress (Krushelnycky et al., 2013); and some coral species, which are known to die at ocean temperatures that are only 0.5-1°C above the maxima experienced prior to 1998 (Hoegh-Guldberg, 1999; Mumby et al., 2007; Pandolfi et al., 2011).

In Category 2 are many marine species whose growth and development are affected by calcium and aragonite concentrations in ocean water, which vary with increasing acidification caused by adding CO_2 to the atmosphere. Already exhibiting detrimental effects are the oyster *Crassostrea gigas* (Barton et al., 2012; Gazeau et al., 2011) in the US Pacific Northwest, where warmer, more acidic waters cause the oyster eggs to die after a few days of apparently normal development. Experimental work, where organisms are reared in waters simulating ocean chemistry expected by the year 2100, also reveals fatal or potentially detrimental effects on other species, including the oysters *Crassostrea virginica* (Miller et al., 2009) and *Pinctada fucata* (Liu et al., 2012), inland silverside fish *Menidia beryllina* (Baumann et al., 2011), Atlantic cod *Gadus morhua* (Frommel et al., 2011), sea bass *Atractoscion nobilis* (Checkley et al., 2009), orange clown fish *Amphiprion percula* (Munday et al., 2009), and damsel fish (*Pomacentrus amboinensis*) (Ferrari et al., 2012). This effect is predicted to be particularly severe for coral-forming species (Ricke et al., 2013). Similarly, many plant populations are stressed by climate change. For example, earlier snowmelt in the Rocky Mountains exposes plants to increased frost damage, (e.g., Inouye, 2008), and declining summer fog causes stress to coastal redwoods (Johnstone and Dawson, 2010). Climate change also causes indirect impacts on plants via outbreaks of pests such as pine bark (Kurz et al., 2008) and spruce bark beetles (National Climate Assessment and Development Advisory Committee, 2013; Bentz et al., 2010).

In Category 3 are many species that will probably experience lethal effects in large parts, but not all, of their geographic ranges. For example, warmer river temperatures could reduce habitat for trout in the Rocky Mountain West up to 50 percent, and locally up to 70 percent, by 2100 (Kinsella et al., 2008). Survival for such species will depend on whether or not viable population sizes will remain in areas where climate does not change unsuitably, and on the potential of surviving individuals to disperse from climatically unsuitable areas into regions with favorable climate. Of particular concern are species now much reduced in numbers of individuals and restricted to protected habitat islands, such as national parks, that are surrounded by human-dominated landscapes where survival of the affected species is not possible without changing societal norms (Early and Sax, 2011).

Plausible vulnerabilities are potentially more severe than the demonstrable vulnerabilities. Of primary concern are probabilities of novel and disappearing combinations of climatic parameters (Williams and Jackson, 2007). Novel climates are those that are created by combinations of temperature, precipitation, seasonality, weather extremes, etc., that exist nowhere on Earth today. Disappearing climates are combinations of climate parameters that will no longer be found anywhere on the planet. Modeling studies suggest that by the year 2100, between 12 percent and 39 percent of the planet will have developed novel climates, and current climates will have disappeared from 10 percent to 48 percent of Earth's surface (Williams et al., 2007). These changes will be most prominent in what are today's most important reservoirs of biodiversity (including the Amazon, discussed in more detail in the "Abrupt Changes in Ecosystems" section above) and if they result in loss of critical aspects of species' ecological niches, a large number of extinctions would result. Other circumstances that have high plausibility of accelerating extinctions include climatically induced loss of keystone species, collateral loss of species not necessarily affected by climate directly but dependent on species removed by climate change (for example, the myriad species dependent on coral-building species, see below), and phenology mismatches (disruption of the links between a species' yearly cycle and the seasons) (Dawson et al., 2011; NRC, 2011a).

Likelihood of Abrupt Changes

It presently is not possible to place exact probabilities on the added contribution of climate change to extinction, but the observations noted above indicate substantial risk that impacts from climate change could, within just a few decades, drop the populations in many species below sustainable levels, which in turn would commit the species to extinction. Thus, even though such species might not totally disappear as a result of climate change within the next two or three decades, climate impacts emplaced during that time would seal the species' fate of extinction over the slightly longer term. On the other hand, the risks of abrupt extinction (within 30 to 80 years) are high for many species that live within two kinds of highly biodiverse ecoystems—tropical and subtropical rainforests such as the Amazon, and coral reefs. Although rainforests presently cover only about 2 percent of Earth's land, they harbor about half of the planet's terrestrial species,[18] and the tropics as a whole contain about two-thirds of all terrestrial animal and plant species (Pimm, 2001). It is these areas that are among those expected to experience the greatest relative difference between 20th century and late 21st century climates, including a large proportion of "disappearing" and "novel" climates (Williams et al., 2007). Coral reefs, which plausibly as a result of climate change could disappear entirely by 2100 and almost certainly will be reduced much in areal extent within the next few decades (Hoegh-Guldberg, 1999; Mumby et al., 2007; Pandolfi et al., 2011; Ricke et al., 2013), are essentially the "rainforests of the sea" (Knowlton and Jackson, 2008) in terms of biodiversity. Coral reefs support 800 hard coral species, over 4,000 fish species, over 25 percent of the world's fish biodiversity, and between 9-12 percent of the world's total fisheries.[19] Species in high-elevation and high-latitude regions may also be especially vulnerable to extinction as their current climate zones disappear.

It is possible to gain some qualitative insights from natural experiments afforded by the fossil record to bound the worst-case scenarios. A 4°C increase in mean global temperature, which is plausible by the year 2100,[20] would make mean global temperature similar to what it was 14 to 15 million years ago (Barnosky et al., 2003). Then, areas that are now at the top of the Continental Divide in Idaho and Montana were occupied by large tortoises that could not withstand freezing temperatures in winters

[18] http://www.nature.org/ourinitiatives/urgentissues/rainforests/rainforests-facts.xml.

[19] http://coralreef.noaa.gov/aboutcorals/values/biodiversity/.

[20] The IPCC AR5 RCP8.5 scenario suggests that exceeding 4.0°C of warming is "about as likely as not and the AR4 suggests warming of 4.0°C by 2100 (relative to 1980-1999) as the 'best estimate' for the A1F1 scenario (IPCC, 2007c; NRC, 2011a). Society may be closer to this trajectory than to the IPCC AR4 A2 scenario, or the AR5 RCP4.5 or 6.0 scenarios. Davis et al., 2013 note that "actual annual emissions have exceeded A2 projections for more than a decade," citing Houghton, 2008 and Boden et al., 2011.

(Barnosky et al., 2007). At the same time, what are now arid lands in Idaho and Oregon supported forests of warm-temperate trees like those in the mahogany family, presently characteristic of central and South America, and in the deserts of Nevada, forests were composed of trees that now are native to the southeastern United States and eastern Asia, for instance, maple, alder, ash, yellowwood, birch, beech, and poplars (Graham, 1999). Given current emissions trajectories, there is a chance that the temperature increase by 2100 could be near 6°C.[21] The last time Earth exhibited a global mean temperature that high, what are now sagebrush grasslands in the southwestern Wyoming and Utah were covered by subtropical, closed canopy forests interspersed with open woodlands (Townsend et al., 2010), reminiscent of subtropical areas in Central America today.

While different continental configurations, elevations, and atmospheric circulation patterns now prevail on Earth, precluding a return to those exact past conditions, the underlying message is that warming of 4°-7° will result in a biotically very different world. At best, changes of such magnitude would trigger dramatic re-organization of ecosystems across the globe that would play out over the next few centuries; at worst, extinction rates would elevate considerably for the many species adapted to pre-global warming conditions, via mechanisms described above (inability to disperse or evolve fast enough to keep pace with the extremely rapid rate of climate change, and disruption of ecological interactions within communities as species respond individualistically).

In the oceans, some insights can be gained by tracking how pH values and relative change in pH values correlated with the most severe past mass extinction event, the end-Permian extinction. At current emissions trends, average pH of the oceans would drop from about 8.1 (current levels) to at least 7.9 in about 100 years (NRC, 2011a).[22] A similar change occurred over the 200,000 years leading up to the end-Permian mass extinction, which resulted in loss of an estimated ~90 percent or more of known species (Chen and Benton, 2012; Knoll et al., 2007). The actual extinction event may have been considerably less than 200,000 years in duration, but the vagaries of geological dating preclude defining a tighter time span. While there may well have been multiple stressors that contributed to end-Permian extinctions, hitting critical thresholds of equatorial warming and acidification are now thought to be major contributors

[21] The IPCC AR4 scenario A1F1 also yields a 66% chance of warming as much as 6.4°C (IPCC, 2007c; NRC, 2011a) and the AR5 scenario a similar chance of warming 5.8°C (IPCC, 2013).

[22] This estimate holds for both the IPCC AR4 A2 Scenario, in which CO_2 concentrations rise to approximately 850 ppm in 2100, or for the the A1F1 Scenario with CO_2 concentrations of around 940 ppm in 2100 (IPCC, 2001), and for the RCP8.5 scenario (IPCC, 2013). NRC (2011a) notes that at 830 ppm, tropical ocean pH would be expected to drop .3 pH units; the A1F1 Scenario would result in a larger decrease in pH.

(Hönisch et al., 2012; Payne and Clapham, 2012; Sun et al., 2012). The end-Permian extinction started from a different continental configuration and global climate, so an exact reproduction is not to be expected, but the potential for a very large number of extinctions in the next few decades, as a result of elevated CO_2 levels that warm the atmosphere and oceans and acidify ocean waters, is analogous (Hönisch et al., 2012; Payne and Clapham, 2012; Sun et al., 2012).

More recently in geological time, the climatic warming at the last glacial-interglacial transition was coincident with the extinction of 72 percent of the large-bodied mammals in North America, and 83 percent of the large-bodied mammals in South America—in total, 76 genera including more than 125 species for the two continents (Barnosky and Lindsey, 2010; Brook and Barnosky, 2012; Koch and Barnosky, 2006). Many of these extinctions occur within and just following the Younger Dryas, and generally they are attributed to an interaction between climatic warming and human impacts (Barnosky et al., 2004; Brook and Barnosky, 2012; Koch and Barnosky, 2006). The magnitude of climatic warming, about 5°C, was about the same as currently-living species are expected to experience within this century, although the end-Pleistocene rate of warming was much slower. Also similar to today, the end-Pleistocene extinction event played out on a landscape where human population sizes began to grow rapidly, and when people began to exert extinction pressures on other large animals (Barnosky, 2008; Brook and Barnosky, 2012; Koch and Barnosky, 2006). The main differences today, with respect to extinction potentials, are that anthropogenic climate change is much more rapid and moving global climate outside the bounds living species evolved in, and the global human population, and the pressures people place on other species, are orders of magnitude higher than was the case at the last glacial-interglacial transition (Barnosky et al., 2012).

Summary and the Way Forward

The current state of scientific knowledge is that there is a plausible risk for climate change to accelerate already-elevated extinction rates, which would result in loss of many more species over the next few decades than would be the case in the absence of climate change. Many of the extinction impacts in the next few decades could be cryptic, that is, reducing populations to below-viable levels, destining the species to extinction even though extinction does not take place until later in the 21st or following century. The losses would have high potential for changing the function of existing ecosystems and degrading ecosystem services (see Chapter 3). The risk of widespread extinctions over the next three to eight decades is high in at least two critically important ecosystems where much of the world's biodiversity is concentrated, tropical/

sub-tropical areas, especially rainforests and coral reefs. The risk of climate-triggered extinctions of species adapted to high, cool elevations and high-latitude conditions also is high.

There are several questions that are still at a nascent stage of discovery:

- Exactly which species in which ecosystems are most at risk?
- Which species extinctions would precipitate inordinately large ecological cascades that would lead to further extinctions?
- What is the impact of climate-induced changes in seasonal timing and species interactions on extinction rates?

Likewise, much remains to be learned about whether loss of biodiversity in all cases means loss of ecosystem services (see Chapter 3, section on Ecosystem Services), and what loss of diversity through extinctions would actually cost humanity.

What can be monitored to see abrupt changes coming? Evaluating trends of species decline/persistence in uniform ways, such as using techniques developed and in place by the International Union for the Conservation of Nature (IUCN), is especially important for non-charismatic species that may be essential in controlling ecosystem function, and for marine invertebrates. In general, it would be useful to monitor species composition, abundance, phenotype, genetic diversity, nutrient cycling, etc. in uniform ways in many different ecosystems, especially those thought to have little impact by humans or otherwise set aside as protected areas (national parks, remote regions, etc.) (Barnosky et al., 2012).

Currently, monitoring is taking place in a variety of contexts, and some data-sharing and uniformity of data sharing is emerging with efforts such as the National Ecological Observatory Network (NEON).[23] But in general different things are being monitored in different ecosystems, and there is little coordination among different groups. Overall, a more uniform, worldwide system of ecosystem/species monitoring is needed (e.g., Pereira et al., 2013). Additionally, a longer time perspective is needed to develop ways to separate the ecological "noise" from the significant ecological signals that would presage biodiversity collapse. This requires comparing changes observed over decades and centuries to long-term ecological baselines of change interpreted from relevant prehistoric records—much as the climate community has done with comparing recent changes with prehistoric proxy data (Barnosky et al., 2012; Hadly and Barnosky, 2009).

Further research is also needed in several key areas:

[23] http://www.neoninc.org/.

1. Developing metrics to set short-term changes observed over decades or centuries in the context of long-term (several hundreds to thousands of years or more) variation in specific ecosystems
2. Better monitoring and modeling of population parameters that would predict extinction risk in a wide variety of species
3. Better understanding of what species are most imperiled by climate change—of those IUCN species in the vulnerable categories, for example, which would be substantially affected by climate change and which would be more resilient
4. Better understanding of which species are true keystones, and which of those are actually at risk from climate
5. Better understanding of how particular life history traits of species predict vulnerability
6. Better predictive models of spatial and demographic responses of species to changes in specific climate parameters
7. Better understanding of the role of species interactions in affecting resilience to climate change
8. Better understanding of the costs—in ecosystem services, economics, and aesthetic/emotional value—of losing species through extinction

With improved understanding of these issues, society can make more informed decisions about potential intervention actions (Figure 2.21).

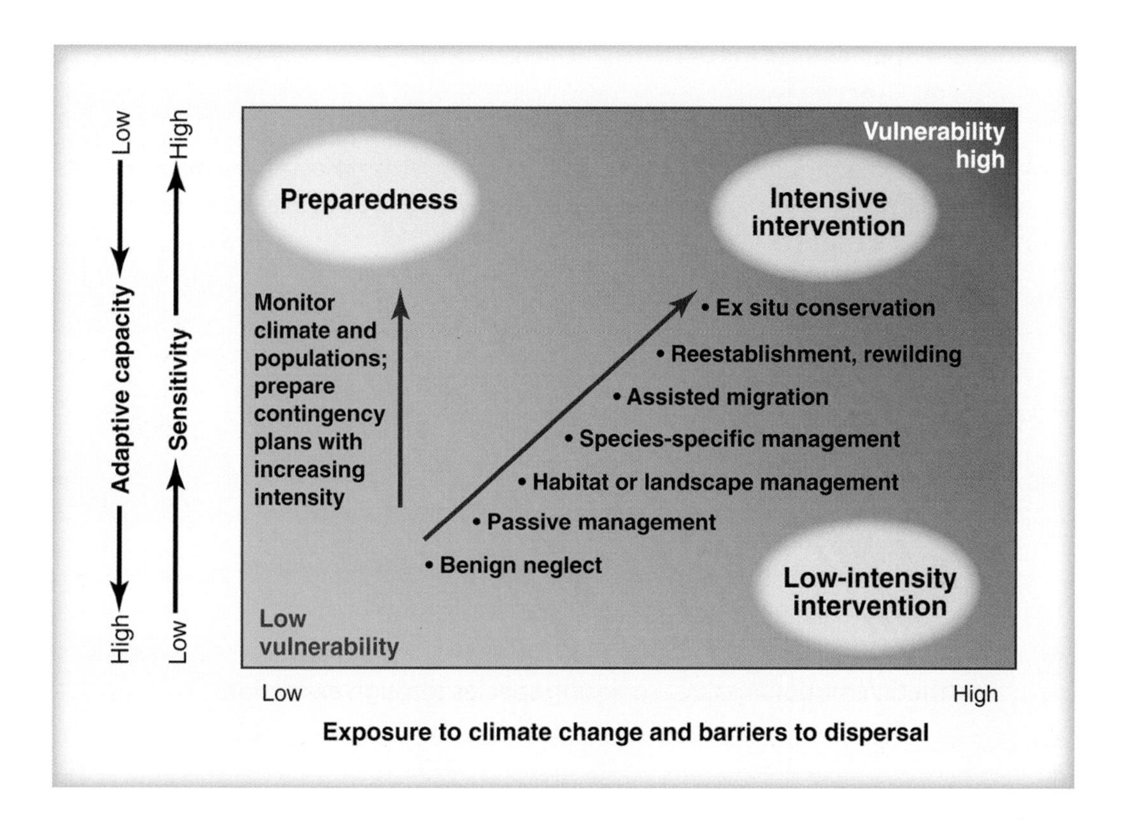

FIGURE 2.21 Improved understanding of adaptive capacity, sensitivity, and exposure to climate change can allow for more informed policy decisions. Potential actions are shown as a function of these variables. Source: Dawson et al., 2011.

Areas of Concern for Humans from Abrupt Changes

Chapter 2 explored a number of potential abrupt climate changes from the point of view of examining the processes. In this chapter, the committee reframes the discussion to look at the issue of abrupt climate changes from the point of view of how they might affect human society. This chapter synthesizes the previous material into how it relates to food security, water security, ecosystem services, infrastructure, human health, and national security.

ECOSYSTEM SERVICES

Abrupt climate impacts may have detrimental effects on ecological resources that are critical to human well-being. Such resources are called "*ecosystem services*" (Box 3.1), which basically are attributes of ecosystems that fulfill the needs of people. For example, healthy diverse ecosystems provide the essential services of moderating weather, regulating the water cycle and delivering clean water, protecting and keeping agricultural soils fertile, pollinating plants (including crops), providing food (particularly seafood), disposing of wastes, providing pharmaceuticals, controlling spread of pathogens, sequestering greenhouse gases from the atmosphere, and providing recreational opportunities (Cardinale et al., 2002; Daily and Ellison, 2002; Daily et al., 2000; Ehrlich et al., 2012; Tercek and Adams, 2013). As explained in Box 3.1, ecosystem services can generally be categorized as provisioning services, regulatory services, cultural services, and supporting services. Currently recognized trajectories of climate change have the potential to cause abrupt changes in each of these categories, in three different ways.

First, gradual changes in the climate system can result in crossing ecologically important threshold values in certain climatic parameters that suddenly cause species to disappear from an area (Chapter 2, Extinctions: Marine and Terrestrial). Examples include soils becoming too dry to support forests, corals dying back because water becomes too warm or acidic, or lizards becoming locally extinct because it is too hot for them to forage (Sinervo et al., 2010). A growing number of studies document that ecosystem transformations that result in loss of biodiversity—as can happen from extinction, or from ecological regime shifts that do not necessarily involve global

BOX 3.1 ECOSYSTEM SERVICES

Humans receive a wide variety of benefits from ecosystem resources and processes (Figure A). The term "ecosystem services" has been used to encapsulate these benefits. Although the notion of human dependence on the services that Earth's ecosystems provide is not new, the definition and categorization of such services were formalized and popularized by the Millennium Ecosystem Assessment in 2005 (MEA, Reid et al., 2005).

The MEA divides ecosystem services into four categories: supporting services, provisioning services, regulating services, and cultural services (Reid et al., 2005):

- *Provisioning Services* are "products obtained from ecosystems," including: food, fuel, freshwater, natural medicines, and pharmaceuticals.
- *Regulating Services* are "benefits obtained from the regulation of ecosystem processes," including: regulation of climate, water, air quality, erosion, pests, and diseases.
- *Cultural Services* are the "nonmaterial benefits people obtain from ecosystems through spiritual enrichment, cognitive development, reflection, recreation, and aesthetic experiences," including: intellectual and spiritual inspiration, ecotourism, and scientific discovery.
- *Supporting services* are ecosystem services required "for the production of all other ecosystem services." These include soil formation, photosynthesis, and nutrient cycling.

FIGURE A Agriculture, specifically corn (A), and fresh seafood (B) are two examples of provisioning services. Regulating services include coastal wetlands, like those of Ashe Island, NC (C), which stabilize shorelines and help buffer against storm erosion, and pollinators such as bees (D), birds, and bats, which are depended upon for thirty-five percent of global crops (Klein et al., 2007). The Catskill Mountains (E) act as a water filtration "plant" for New York City—another example of a regulating service. SOURCE: (A) NSF, (B) The George F. Landegger Collection of District of Columbia Photographs in Carol M. Highsmith's America, Library of Congress, Prints and Photographs Division, (C) NOAA, (D) USDA/Stephen Ausmus, (E) FWS.

BOX 3.2 DETRIMENTAL EFFECTS OF BIODIVERSITY LOSS IN GENERAL

An emerging body of literature is documenting that reducing biodiversity is detrimental to ecological function, and ultimately to ecosystem services that benefit humans, in at least six ways (Cardinale et al., 2012; Collen et al., 2012; WRI, 2005) (Box 2.3). The following list of the costs of losing biodiversity is extracted from the review by Cardinale et al. (2012), as based on their vetting of published studies.

1. *"Reductions in the number of genes, species and functional groups of organisms reduce the efficiency by which whole communities capture biologically essential resources (nutrients, water, light, prey), and convert those resources into biomass."* This implies that biodiversity-poor systems require more resources to be input by humans in order to maintain them, for example, addition of artificial fertilizers to maintain productivity.
2. *"There is mounting evidence that biodiversity increases the stability of ecosystem functions through time."* This means that biodiverse systems are more dependable in providing benefits to humanity, for example, carbon sequestration in forests, because they better withstand unanticipated perturbations.
3. *"The impact of biodiversity on any single ecosystem process is nonlinear and saturating, such that change accelerates as biodiversity loss increases."* Therefore, as biodiversity loss increases, ecological systems are increasingly likely to suddenly hit a "tipping point," for example, economically important fisheries in marine coastal areas being replaced by systems dominated by algae and jellyfish.
4. *"Diverse communities are more productive because they contain key species that have a large influence on productivity."* This means that removing a single species can have unexpectedly large effects in transforming an entire ecosystem, such as removal of elephants changing a grassland savannah into a forest.
5. *"Loss of diversity across trophic levels has the potential to influence ecosystem functions even more strongly than diversity loss within trophic levels."* Loss of diversity is seldom restricted to the trophic level that experiences the loss because the connections between interacting species are altered. For example, loss of top predators has been shown to reduce plant biomass significantly.
6. *"Functional traits of organisms have large impacts on the magnitude of ecosystem functions, which give rise to a wide range of plausible impacts of extinction on ecosystem function."* This means that removal of species with unique ecological attributes is especially disruptive to ecosystems.

Quoted from Cardinale et al., 2012, pg. 59.

---

extinction but cause extirpation of species locally or regionally—detrimentally affect ecosystem services (Box 3.2). Second, extreme events (increased frequency of floods or drought, for instance) can trigger sudden, regional catastrophes that wipe out natural ecosystems (coastal salt marshes, for instance) or human-controlled ones (crop fields). Third, cascading effects of abrupt, climatically-triggered changes to the physical environment can cause abrupt transformation in widespread ecosystems, such as loss of sea ice affecting the marine food chain, or loss of coral reefs impacting fisheries (discussed and cited above).

Food

The spectrum of abrupt disruptions to ecosystem services as a result of climate change is broad, but of particular concern are those that would impact essential provisioning services such as food production and water availability. The potential disruptions of food (and water, see below) supplies could be one of the most serious manifestations of abrupt climate change, especially when put in the context of the changing global food system.

The world currently has a population exceeding 7 billion (and is likely to grow to ~9 billion by 2050) and an estimated ~850 million are already considered food insecure (Godfray et al., 2010). The challenges of food security today are driven mainly by poverty, the lack of access to food, and poor institutions (Godfray et al., 2010). Even a slight added impact from climate change, therefore, could lead to significant, abrupt, and problematic food shortages.

Tilman et al. (2011) have estimated that by 2050, world demand for agricultural products will increase by roughly 100 percent, largely driven by rising incomes and increasing meat consumption. This challenge is made more cogent by realizing that even without potential impacts of climate change, providing food security, meeting growing demands for agricultural products, and ensuring the environmental sustainability of agricultural systems worldwide will require a multi-faceted approach (Foley et al., 2011; Godfray et al., 2010; Tilman et al., 2011). Such an approach will need to concentrate on boosting yields (especially in places where yields are low today), improving the resource efficiency of agriculture (especially the water, nutrients, and energy used per calorie of food delivered), avoiding further deforestation and land degradation, shifting diets and biofuels to more sustainable trajectories, and reducing food waste across the entire supply chain (Foley et al., 2011).

To date, investigations of how climate change will affect crop production and food systems have mainly focused on long-term changes in the mean climate (e.g., annual rainfall, patterns of temperature). Mainly, these studies have split into two broad categories: those that do explicitly consider the adaptation of farmers to climate change, and those that do not.

One of the first studies to consider the impacts of climate change on agriculture was conducted by Rosenzweig and Parry (1994). Adaptations include changes in planting date, variety, and crop, as well as changes in applications of irrigation and fertilizer. They found that climate change scenarios (for 2060) *with* adaptation could result in increasing yields (from +3 to +10 percent) in developed countries, while developing countries would see decreasing yields of approximately 6 percent. However, with no adaptation, developed countries would see production changes from –4 percent to +11 percent, and developing country yields would see decreases of 9 to 12 percent. Parry et al. (2004) presented an updated version of this work for a variety of climate change scenarios for 2020, 2050, and 2080, and made similar conclusions: yield increases in developed countries will tend to counteract decreases in developing countries. Typical increases in production for the developed world range from 3 to 8 percent, and typical decreases for the developing world range from 2 to 7 percent.

More recently, Deryng et al (2011) used the PEGASUS process-based crop model and found that global maize production for 2050, under a climate change scenario based on rapid economic growth (A1B; see IPCC, 2007), changes by –15 percent, and under a scenario based on more modest economic growth (B1) changes by -8 percent, if farm-level adaptation (especially changing planting dates) is taken into account. However, without farm-level adaptation of planting dates, the yields decreases are estimated 30 percent and 20 percent respectively. A new study[1] uses a cross-sectional method based on the shifting climate zones and find that global maize production in 2050 (under and A1B scenario) could decrease by 7 percent and under a B1 scenario will decrease by 3.5 percent.

In short, it is clear that changes in climate will have profound impacts on global food production and, in turn, food security (Easterling et al., 2012; Lobell and Gourdji, 2012; Lobell et al., 2011). Modeling and statistical analyses consistently show that climate change could introduce substantial changes to global food production (some positive, many strongly negative). However, the exact magnitude of these changes depends on the assumptions made about the adaptation of farmers to climate change (Easterling et al., 2012). This presents a particular challenge in the face of abrupt climate change (compared to slower changes in climate that may occur over many decades). Will abrupt changes in climate cause more severe dislocations in agriculture, because it leaves less time for farmers and agricultural markets to adapt? This remains a critical area for future research.

[1] Personal communication, J.S. Gerber, 2013.

Water

Humans currently withdraw roughly 4,000 km^3 of water globally, mainly for irrigation (~70 percent), industry (~20 percent), and domestic use (~10 percent). Water consumption (the net use of water from a watershed, accounting for water return flows and recycling within the same watershed) globally is estimated to be 3,000 km^3, with agricultural irrigation taking an even larger share (~90 percent) (World Water Council, 2000). Therefore, climate changes, mainly through changes in precipitation and evapotranspiration over watersheds that people depend upon, can cause serious, abrupt (yearly to decadal) changes to critical freshwater resources. Additional concerns about freshwater resources are linked to snow and ice melt, which provides critical drinking and irrigation water to many people worldwide, including highly populous and/or politically sensitive areas such as Pakistan, India, and along the border of China and Nepal.

As with other ecosystem services, it is necessary to interpret potential climatic impacts on freshwater resources within the broader context of how water resources are already stressed around the world. Freshwater resources are already reaching limits under the increasing demands of a growing population, rising incomes, and increasing per capita consumption (particularly through food). Vorosmarty et al. (2000), for example, demonstrated that changes to the current patterns from water consumption and withdrawals already exceed the expected changes to the water cycle anticipated from climate change. Furthermore, increasing demands on water (estimated from population growth and economic development) will greatly exceed expected changes from climate change. Vorosmarty et al. summarize the situation as, "We conclude that impending global-scale changes in population and economic development over the next 25 years will dictate the future relation between water supply and demand to a much greater degree than will changes in *mean* climate." (emphasis added)

Groundwater aquifers, for example, are being depleted in many parts of the world, including the southeast of the United States. Groundwater is critical for farmers to ride out droughts, and if that safety net reaches an abrupt end, the impact of droughts on the food supply will be even larger. Satellites measuring gravity now reveal that groundwater supplies have decreased rapidly around the world over the past decade, including key aquifers in California, the High Plains, and the southeastern United States (Famiglietti et al., 2011). Groundwater is a key part of successful adaptation to periodic drought, which in turn is a key aspect of maintaining stable food supplies. In many cases it is unknown how long this situation could continue without water availability reaching an end, possibly an abrupt end, although history is clear in showing that groundwater supplies can indeed be depleted, parts of the Ogallala Aquifer (under the US Great Plains) being an example. Questions remain about the future of this potential abrupt change, but the potential impact, especially on national and global food supplies, is substantial.

On a larger scale, changes in atmospheric circulation (e.g., changes to monsoon circulations), precipitation variability (e.g., more high extreme rainfalls) and abrupt changes in the condition of snow and ice packs have high potential of reducing crop productivity in some areas, and raising it in others—such shifts will have downstream impacts on local and national economies. For example, largely due to water-delivery issues related to climate change, cereal crop production is expected to fall in areas that now have the highest population density and/or the most undernourished people, notably most of Africa and India (Dow and Downing, 2007). In the United States, key crop-growing areas, such as California, which provides half of the fruits, nuts, and vegetables for the United States, will experience uneven effects across crops, requiring farmers to adapt rapidly to changing what they plant (Kahrl and Roland-Holst, 2012; Lobell et al., 2006).

Fisheries

Degradation of coral reefs by ocean warming and acidification will negatively affect fisheries, because reefs are required as habitat for many important food species, especially in poor parts of the world. For example, in the poorest countries of Africa and south Asia, fisheries largely associated with coral reefs provide more than half of the protein and mineral intake for more than 400 million people (Hughes et al., 2012). On a broader scale, many fisheries around the world can be expected to experience changes as ocean temperatures, acidity, and currents change (Allison et al., 2009; Jansen et al., 2012; Powell and Xu, 2012), with attendant socio-economic impacts (Pinsky and Fogarty, 2012). One study suggests climate change, combined with other pressures on fisheries, may result in a 30–60 percent reduction in fish production by 2050 in areas such as the eastern Indo-Pacific, and those areas fed by the northern Humboldt and the North Canary Currents (Blanchard et al., 2012). Because other pressures, notably over-fishing, already stress fisheries, a small climatic stressor can contribute strongly to hastening collapse (Hidalgo et al., 2012).

Other Provisioning Services

Outside the food and water sector, abrupt changes to other provisioning services also are very likely as a result of in-progress climate change (Reid et al., 2005, see Box 3.1).

Forest diebacks (Anderegg et al., 2013) and reduced tree biodiversity (Cardinale et al., 2012) can be expected to have major impacts on timber production. Such is already the case for millions of square miles of beetle-killed forests throughout the American West. Drought-enhanced desertification of dryland ecosystems may cause famines and migrations of environmental refugees (D'Odorico et al., 2013).

In several documented cases the efficacy of provisioning services correlates positively with the biodiversity of an ecosystem (Cardinale et al., 2012); thus, the loss of biodiversity through climate-caused or climate-exacerbated extinctions is of considerable concern. Among the provisioning services that have been shown to increase with biodiversity are: intraspecific genetic diversity increasing the yield of commercial crops; tree species diversity enhancing production of wood in plantations; plant species diversity in grasslands improving the production of fodder; higher diversity of fish leading to greater stability of fisheries yields; higher plant diversity increasing resistance to invasion by less-desirable exotic species, and in decreasing prevalence of fungal and viral infections (Cardinale et al., 2012). Some studies suggest that increased biodiversity also increases the following ecosystem services: carbon storage, pest reduction, reduction in animal diseases, fisheries yields, flood protection, and water quality. However, the efficacy of higher biodiversity in promoting these services still is under study, with conflicting results for different studies clouding the generality of the relationship (Cardinale et al., 2012).

Regulatory Services

Also of concern is the potential loss of regulatory services, which buffer the effects of environmental change (Reid et al., 2005). For example, tropical forest ecosystems slow the rate of global warming both by absorbing atmospheric carbon dioxide and through latent heat flux (Anderson-Teixeira et al., 2012). Coastal saltmarsh and mangrove wetlands buffer shorelines against storm surge and wave damage (Gedan et al., 2011). Grassland biodiversity stabilizes ecosystem productivity in response to climate variation (see Cardinale et al., 2012 and references therein). Climate change has the clear potential to exacerbate losses of these critical ecosystem services (for instance, decrease in rainforests, desertification) and attendant impacts on human societies.

Direct Economic Impacts

Some species currently at risk of extinction, and some of those which will be further imperiled by ongoing climate change, provide significant economic benefits to people who live in the surrounding areas, as well as significant aesthetic and emotional benefits to millions of others, primarily through ecotourism, hunting, and fishing. At the international level, for example, ecotourism—largely to view elephants, lions, cheetahs, and other threatened species—supplies around 14 percent of Kenya's GDP as of 2013 (USAID, 2013) and supplied 13 percent of Tanzania's in 2001 (Honey, 2008). Yet in a single year, 2009, an extreme drought decimated the elephant population and populations of many other large animals in Amboseli Park, Kenya. Increased frequency of such extreme weather events could erode the ecotourism base on which the local economies depend. Other international examples include ecotourism in the Galapagos Islands—driven in a large part to view unique, threatened species—which contributed 68 percent of the 78 percent growth in GDP of the Galapagos that took place from 1999–2005 (Taylor et al., 2008).

Within the United States, direct economic benefits of ecosystem services also are substantial; for example, commercial fisheries provide approximately one million jobs and $32 billion in income nationally (NOAA, 2013). Ecotourism also generates substantial revenues and jobs in the United States—visitors to national parks added $31 billion to the national economy and supported more than 258,000 jobs in 2010 (Stynes, 2011). For Yellowstone National Park, which attracts a substantial number of visitors for wildlife viewing, visitors in 2010 contributed $334 million to the local economies, and created 4,900 local jobs (Stynes, 2011). Wildlife in Yellowstone is undergoing substantial changes, as evidenced by the clear amphibian decline as a result of drying up of breeding ponds (McMenamin et al., 2008). Visitors to Yosemite National Park in 2009 created 4,597 jobs, and yielded $408 million in sales revenues, $130 million in labor income, and $226 million in value added (Cook, 2011). Recent work there demonstrates that many of the small mammals are shifting their geographic ranges, with as yet unknown consequences to the overall ecosystem, as a result of climate change over the last century (Moritz et al., 2008).

INFRASTRUCTURE

The built environment is at risk from abrupt change. Examples near coasts include infrastructure such as roads, power lines, sewage treatment plants, and subway systems located close enough to the ocean and at a low enough elevation to be subject to the direct and indirect (e.g., storm surges) impacts of sea level rise. Other examples from northern latitudes include roads built on permafrost in Alaska, where that permafrost is now melting causing the roads to buckle and heave. Less obviously, there are also systems whose useful lifetimes are cut short by gradual changes in baseline climate. Such systems are experiencing abrupt impacts if they are built to last a certain period of time, and priced such that they can be amortized over that lifetime, but their actual lifetime is artificially shortened by climate change. One example would be a large air conditioning system for computer server rooms. If maximum high temperatures rise faster than planned for, the lifetime of such systems would be cut short, and new systems would need to be installed at added cost to the owner of the servers. Another example is storm runoff drains in cities and towns. These systems are sized to handle large storms that precipitate a certain amount of water in a certain period of time. Rare storms, such as a 1000-year event, are typically not considered when choosing the size of pipes and drains, but the largest storms that occur annually up to once per decade or so are considered. As the atmosphere warms and can hold more moisture, the amount of rain per event is increasing (Westra et al., 2013), changing the baseline used to size storm runoff systems, and thus their utility, generally long before the systems are considered to have reached their useful lifetimes.

Another type of infrastructure problem associated with abrupt change is the infrastructure that does not exist, but will need to after an abrupt change. The most glaring example today is the lack of US infrastructure in the Arctic as the Arctic Ocean becomes more and more ice free in the summer. For example, the United States lacks sufficient ice breakers that can patrol waters that, while seasonally open in many places, will still have extensive wintertime ice cover. Servicing and protecting our activities in this resource-rich region is now a challenge, one that only recently, and abruptly, emerged. This challenge has illustrated a time scale issue associated with abrupt change. Currently, it will take years to rebuild our fleet of ice-breakers, but because of the rapid loss of sea ice in 2007 and more recently, the need for these ships is now (NRC, 2007; O'Rourke, 2013).

Coastal Infrastructure

Globally, about 40 percent of the world's population lives within 100 km of the world's coasts. While complete inventories are lacking, the accompanying infrastructure—from the obvious, such as roads and buildings, to the less obvious but no less critical, such as underground services (e.g., natural gas and electric lines)—is easily valued in the trillions of dollars, and this does not include ecosystem services such as fresh water supplies, which are threatened as sea level rises. A nearly equal percentage of the US population lives in Coastal Shoreline Counties.[2] In addition, coastal counties are more densely populated than inland ones. The National Coastal Population Report, Population Trends from 1970 to 2020 (NOAA, 2013), reports that coastal county population

[2] http://stateofthecoast.noaa.gov/.

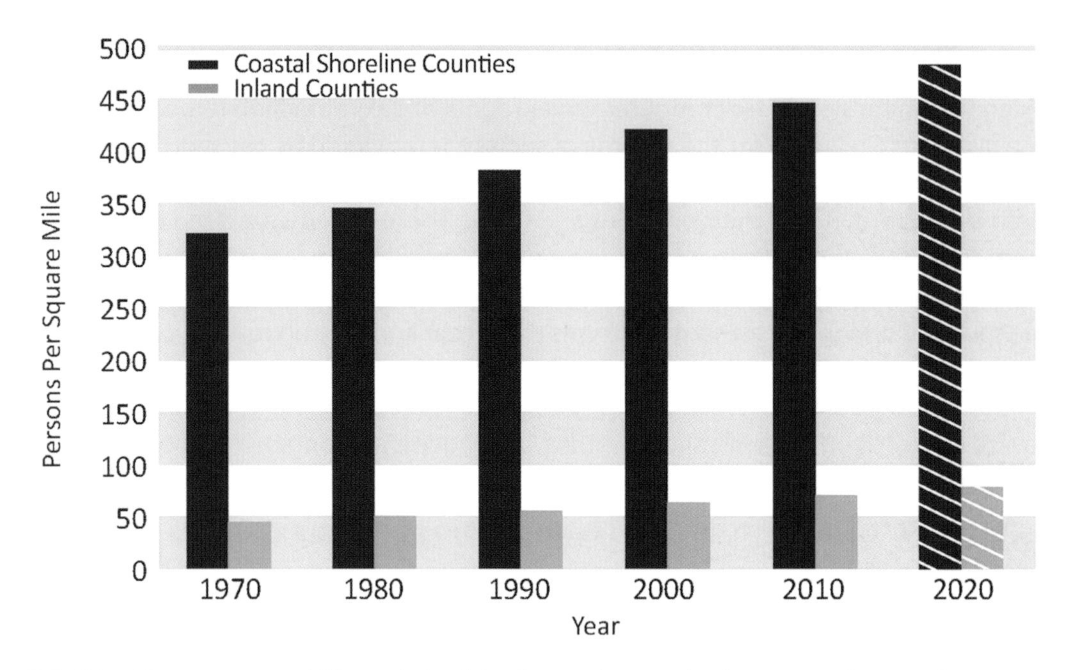

FIGURE 3.1 The percentage of US population living in near the coast has been increasing over the past several decades. In 2010, 39 percent of the US population lived in Coastal Shoreline Counties (less than 10 percent of the total land area excluding Alaska). The population density of Coastal Shoreline Counties is over six times greater than the corresponding inland counties. Source: NOAA, 2013.

density is over six times that of inland counties (Figure 3.1). Consequently, the United States has a large amount of physical assets located near coasts and currently vulnerable to sea level rise and storm surges exacerbated by rising seas (See Chapter 2 and especially Box 2.1 for additional discussion of this issue.) For example, the National Flood Insurance Program (NFIP) currently has insured assets of $527 billion in the coastal floodplains of the United States, areas that are vulnerable to sea level rise and storm surges. Examples of significant payouts include the costs of Hurricane Katrina, which totaled $16 billion from NFIP, and significantly more than that for private insurers, and the recent costs of Superstorm Sandy (Figure 3.2), which are still being totaled, but which will likely exceed Katrina by a large amount. In addition, nearly half of the US gross domestic product, or GDP, was generated in the Coastal Shoreline Counties along the oceans and Great Lakes (see NOAA State of the Coast[3]). Despite the ongoing rise of sea level, and the frequent, high-profile illustrations of the value and vulnerabil-

[3] http://stateofthecoast.noaa.gov/.

FIGURE 3.2 During Superstorm Sandy in 2012, storm surges brought water inland and flooded subway terminals in the New York area—part of the billions of dollars in damages from that storm. SOURCE: Port Authority of New York and New Jersey.

ity of coastal assets at risk, there is no systematic, ongoing, and updated cataloging of coastal assets that are in harm's way as sea level rises. Overall, there is a need to shift to more holistic planning, investment, and operation for global sea ports (Becker et al., 2013).

Arctic Transportation and Infrastructure

Some of the most apparent infrastructure impacts are in the Arctic, owing to both the rapidity of summer sea ice loss in the Arctic Ocean and the non-linear rise of air temperatures there relative to the global mean ("Arctic climate amplification"). For human transportation systems, these trends have both positive and negative impacts, with rising maritime access in seasonally frozen rivers and seas but declining overland access to seasonally frozen ground (Smith, 2010; Stephenson et al., 2011).

Permafrost, or permanently frozen ground, is ubiquitous around the Arctic and sub-Arctic latitudes and the continental interiors of eastern Siberia and Canada, the Tibetan Plateau and alpine areas. As such, it is a substrate upon which numerous pipelines, buildings, roads and other infrastructure have (or could be) built, so long as these structures are properly designed to not thaw the underlying permafrost. For areas underlain by ice-rich permafrost, severe damage to permanent infrastructure can result from settlement of the ground surface as the permafrost thaws (Nelson et al., 2001, 2002; Streletskiy et al., 2012). These terrestrial problems are driven by lessened ground freeze owing to milder winters and/or deeper snowfall (which insulates the ground) that are hallmarks of the Arctic climate amplification.

Numerous engineering problems are associated with thawing of ground permafrost, including loss of soil bearing strength, increased soil permeability, and increased potential for thermokarsting, differential thaw settlement, and heave (Shiklomanov and Streletskiy, 2013). Over the past 40 years, significant losses (>20 percent) in ground load-bearing capacity have been computed for large Arctic population and industrial centers, with the largest decrease to date observed in the Russian city of Nadym where bearing capacity has fallen by more than 40 percent (Streletskiy et al., 2012). Numerous structures have become unsafe in Siberian cities, where the percentage of dangerous buildings ranges from at least 10 percent to as high as 80 percent of building stock in Norilsk, Dikson, Amderma, Pevek, Dudina, Tiksi, Magadan, Chita, and Vorkuta (ACIA, 2005). Problems are also apparent on the Tibetan Plateau, where mean annual ground temperatures have risen as much as 0.5°C in the past 30 years with damages to built infrastructure caused by thaw settlement and slumping in the affected regions (Yang et al., 2010).

The second way in which milder winters and/or deeper snowfall reduce human access to cold landscapes is through reduced viability of winter roads (also called ice roads, snow roads, seasonal roads, or temporary roads). Like permafrost, winter roads are negatively impacted by milder winters and/or deeper snowfall (Hinzman et al., 2005; Prowse et al., 2011). However, the geographic range of their use is much larger, extending to seasonally frozen land and water surfaces well south of the permafrost limit. They are most important in Alaska, Canada, Russia, and Sweden, but also used to a lesser extent (mainly river and lake crossings) in Finland, Estonia, Norway, and the northern US states. These are seasonal features, used only in winter when the ground and/or water surfaces freeze sufficiently hard to support a given vehicular weight. They are critically important for trucking, construction, resource exploration, community resupply and other human activities in remote areas. Because the construction cost to build a winter road is <1 percent that of a permanent road (e.g., ~$1300/km versus $0.5–1M/km, Smith, 2010) winter roads enable commercial activity in remote northern areas that would otherwise be uneconomic.

Since the 1970s, winter road season lengths on the Alaskan North Slope have declined from more than 200 days/year to just over 100 days/year (Hinzman et al., 2005). Based on climate model projections, the world's eight Arctic countries are all projected to lose significant land areas (losses of 11 percent 82 percent) currently possessing climates suitable for winter road construction (Figure 3.3), with Canada (400,000km^2) and Russia (618,000km^2) experiencing the greatest losses in absolute land area terms (Stephenson et al., 2011).

Figure 3.3 also presents a first attempt to quantify navigation potential for ships. In the Arctic Ocean, climate model projections of thinning sea ice thickness, lower sea ice concentration, lower multi-year ice (MYI) fraction, and shorter ice-covered season all enable increased accessibility to ships. Using the CCSM4 climate model, Stephenson et al. (2011; 2013) quantified these trends for three different ship classes (Polar Class 3, Polar Class 6, and common open-water ships) from present-day to the late 21st century. In general, the Russian Federation is projected to experience the greatest increase (both in percent change and total marine-accessible area) in accessibility to its offshore Exclusive Economic Zone (EEZ), followed by Greenland and Norway. Offshore accessibility increases for Canada and the United States are projected to be less than for the Russian EEZ, owing to greater ice persistence in the Canadian Archipelago and already high accessibility off the North Slope of Alaska today. The timing and magnitude of these projected marine accessibility increases are likely conservative, both because most GCM projections of sea ice loss generally lag behind observations and the CCSM4 model in particular has weaker Arctic climate amplification than previous versions (e.g., ~16 percent less than CCSM3, despite higher global warming; Vavrus et al., 2012). When compared to other GCMs, the CCSM4 model also tends to project greater sea ice cover throughout the 21st century relative to other models (Massonnet et al., 2012).

A second impact of declining sea ice thickness and concentration is decreased shipping distance and travel time through summer trans-polar routes linking the Atlantic and Pacific oceans (Figure 3.4). The shipping distance between Shanghai and Rotterdam, for example, is approximately ~19,600 and ~25,600 km, respectively via the Suez or Panama canals, but only ~15,800 km over the northern coast of Russia (the Northern Sea route) or ~17,600 km through the Canadian archipelago (the Northwest Passage). Although the prospect of such trans-Arctic routes materializing has attracted considerable media attention (and indeed, 46 vessels transited the Northern Sea Route during the 2012 season), it is important to point out that these routes would

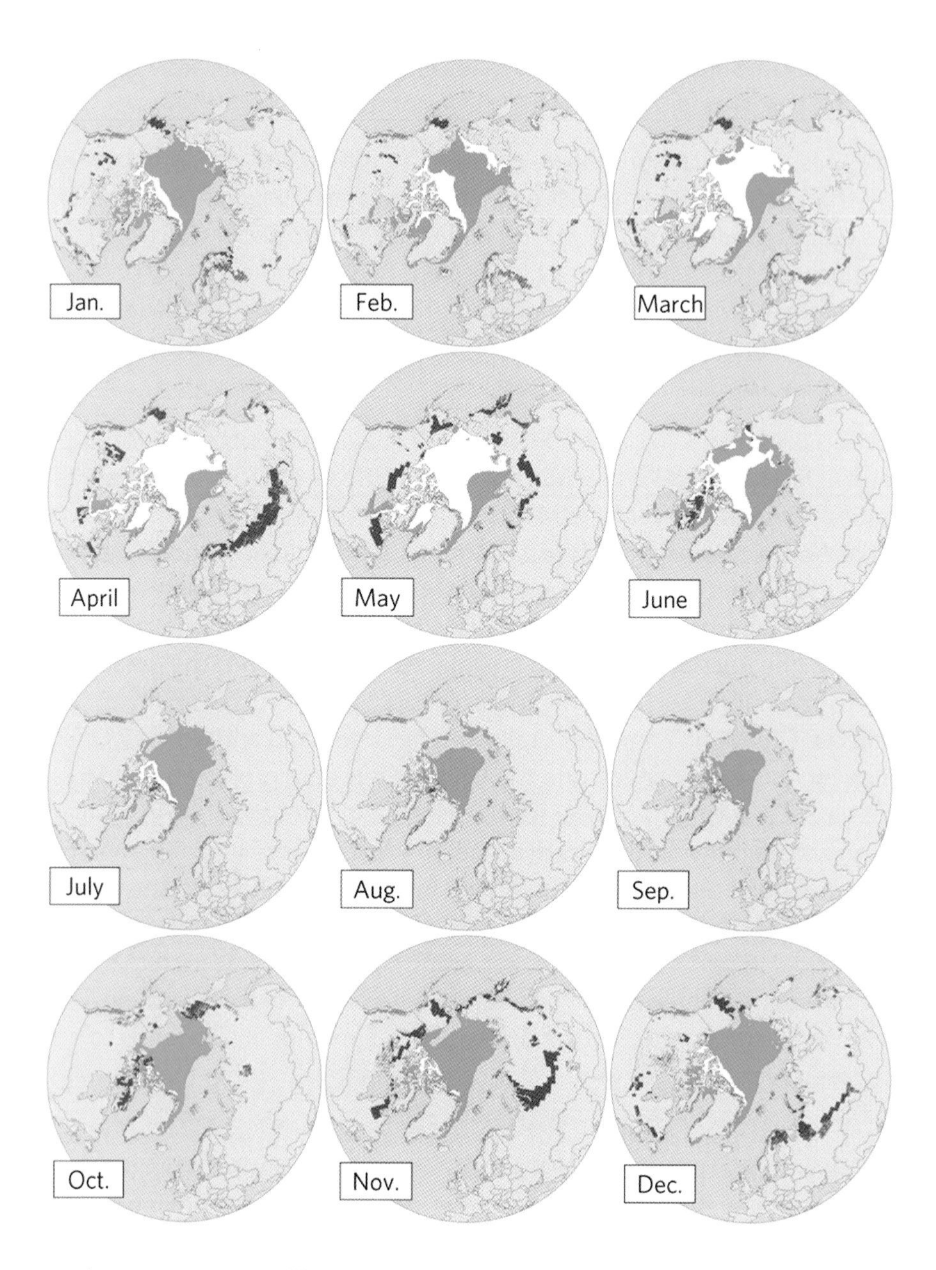

FIGURE 3.3 Changes in marine and land-based transportation accessibility by midcentury, calculated by subtracting midcentury (2045-2059) from baseline (2000-2014) conditions. Green indicates where new maritime access to moderately ice-strengthened ships (Canadian Type A icebreaker) will become enabled. Red indicates where conditions presently suitable for building temporary winter roads (assuming 2000 kg weight vehicles) will be lost. All eight Arctic states are projected to suffer steep declines (11 to 82 percent) in winter road potential, caused by by milder winters and deeper snow accumulation (from Stephenson et al., 2011).

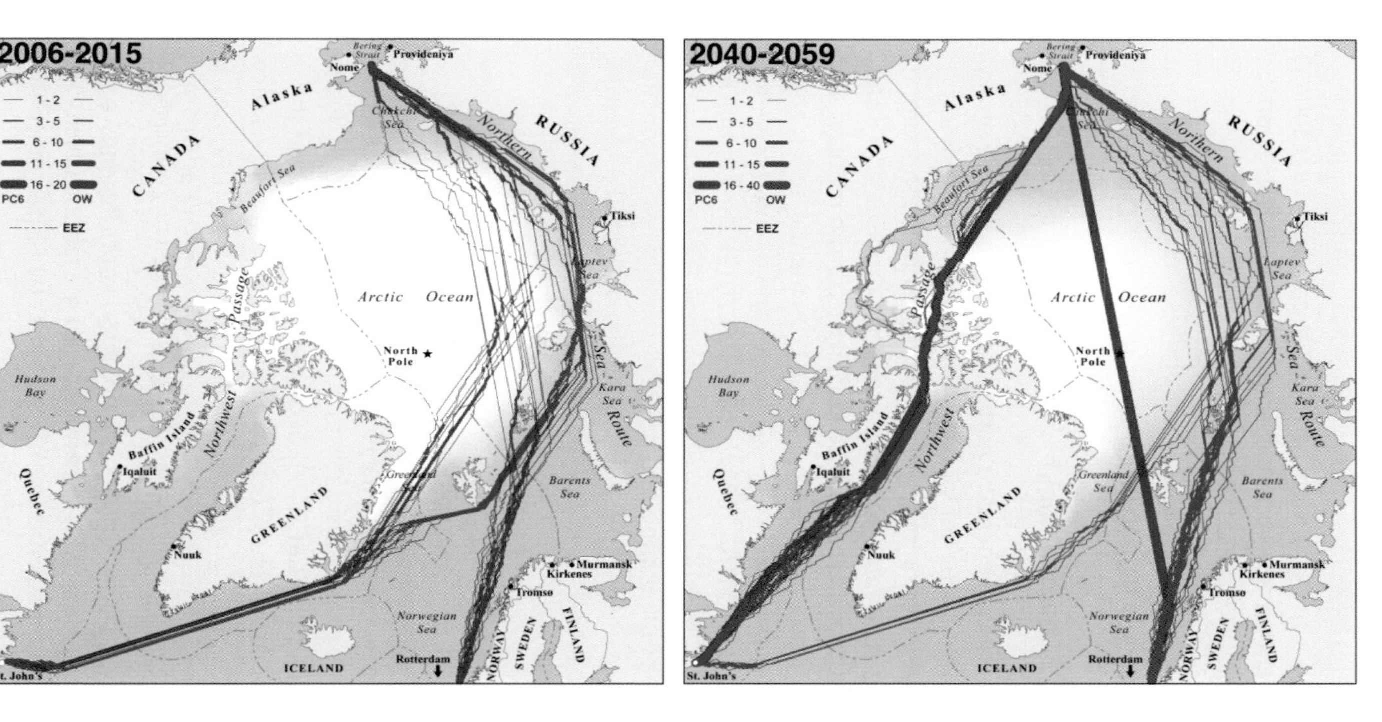

FIGURE 3.4 Fastest Trans-Arctic navigation routes during the peak shipping month of September at present (Septembers 2006-2015) and by midcentury (Septembers 2040-2059) as driven by ensemble-averaged GCM projections of sea ice concentration and thickness (from the ACCESS1.0, ACCESS1.3, GFDL-CM3, HadGEM2-CC, IPSL-CM5A-MR, MPI-ESM-MR, CCSM4 models), for hypothetical ships seeking to cross the Arctic Ocean between the North Atlantic (Rotterdam, The Netherlands and St. Johns, Newfoundland) and the Pacific (Bering Strait). Red lines indicate fastest available routes for Polar Class 6 icebreakers; blue lines indicate fastest available routes for common open-water ships. Where overlap occurs, line weights indicate the number of successful transits using the same navigation route. These particular simulations assume a "medium-low" increase in greenhouse warming (+4.5 Watts/m^2 increase in radiative forcing, called the RCP 4.5 scenario), further simulations assuming a "high" increase in greenhouse warming (+8.5 Watts/m^2, the RCP 8.5 scenario) show further increases in navigability. Dashed lines indicate national 200-nautical mile Exclusive Economic Zone (EEZ) boundaries; white backdrops indicate period-averaged sea ice concentrations (figure adapted from Smith and Stephenson, 2013).

operate only in summer, and numerous other non-climatic factors remain to discourage trans-Arctic shipping including lack of services, infrastructure, and navigation control, poor charts, high insurance and escort costs, unknown competitive response of the Suez and Panama Canals, and other economic factors (AMSA, 2009; Liu and Kronbak, 2010; Brigham, 2010, 2011).

OTHER AREAS OF IMPORTANCE FOR HUMANS FROM ABRUPT CHANGES

Human Health

There are a number of potential adverse effects to human health that may be brought on by changes in the climate. Related issues of food and water security have been discussed in previous sections. This section briefly describes several other human health-related impacts—heat waves, vector-borne and zoonotic diseases, and water-borne diseases—but there are others, including potential impacts from reduced air quality, impacts on human health and development, impacts on mental health and stress-related disorders, and impacts on neurological diseases and disorders (see for example Portier et al., 2010; NRC, 2001; WHO, 2000; WHO/WMO, 2012). The committee stresses that this brief discussion is intended to make the point that human health issues are in many ways tied to abrupt change, and its brevity should not be construed as an indication of the importance of the topic. A full treatment of this subject would be much more extensive, but is beyond the scope of this study as well as the expertise of this committee.

Heat waves cause heat exhaustion, heat cramps, and heat stroke; heat waves are one of the most common causes of weather-related deaths in United States (USGCRP, 2009). Summertime heat waves will likely become longer, more frequent, more severe, and more relentless with decreased potential to cool down at night. Increases in heat-related deaths due to climate change are likely to outweigh decreases in deaths from cold snaps (Åström et al., 2013; USGCRP, 2009). In general, heat waves and the associated health issues disproportionately affect more vulnerable populations such as the elderly, children, those with existing cardiovascular and respiratory diseases, and those who are economically disadvantaged or socially isolated (Portier et al., 2010). Increasing temperature and humidity levels can cross thresholds where it is unsafe for individuals to perform heavy labor (below a direct physiological limit). Recent work has shown that environmental heat stress has already reduced the labor capacity in the tropics and mid-latitudes during peak months of heat stress by 10 percent, and another 10 percent decrease is projected by 2050 (Dunne et al., 2013) with much larger decreases further into the future.

Heavy rainfall and flooding can enhance the spread of water-borne parasites and bacteria, potentially spreading diseases such as cholera, polio, Guinea worm, and schistosomiasis. "Outbreaks of waterborne diseases often occur after a severe precipitation event (rainfall, snowfall). Because climate change increases the severity and frequency of some major precipitation events, communities—especially in the developing world—could be faced with elevated disease burden from waterborne diseases" (Portier et al., 2010). Individual extreme events (see section on Climate Extremes in Chapter 2) could result in abrupt changes in the spread of these diseases, but overall, the impact of climate change on these diseases is not well established.

Vector-borne diseases are those in which an organism carries a pathogen from one host to another. The carrier is often an insect, tick, or mite, and well-known examples include malaria, yellow fever, dengue, murine typhus, West Nile virus, and Lyme disease. Zoonotic diseases are those that are transmitted from animals to humans by either contact with the animals or through vectors that carry zoonotic pathogens from animals to humans; examples include Avian Flu, and H1N1 (swine flu). Changes in climate may shift the geographic ranges of carriers of some diseases. For example, the geographic range of ticks that carry Lyme disease is limited by temperature. As air temperatures rise, the range of these ticks is likely to continue to expand northward (Confalonieri et al., 2007). Overall, the spread of vector-borne and zoonotic diseases that are climate-sensitive will depend heavily on both climate and non-climate factors.

The question for this report is whether any of these effects on human health are likely to change abruptly in the coming decades. One can imagine a gradual migration of insect species over decades, or abrupt outbreaks of waterborne diseases triggered by extreme weather events like floods. Health impacts have the potential to increase the costs and the abruptness of the human health impacts of climate change.

National Security

The topic of climate and national security has been discussed elsewhere (see for example Busby, 2007; Fingar, 2008; McElroy and Baker, 2012; Sullivan et al., 2007; Youngblut, 2010), including a recent review entitled *Climate and Social Stresses: Implications for Security Analysis* (NRC, 2012b). Consequently, remarks here on the subject will be brief, but as with health issues (above), brevity should not be interpreted as an indication of importance. The topic is of vital concern, and interested readers are directed to the aforementioned NRC report and references therein, as well as the excellent discussion of this topic by Schwartz and Randall (2003).

Overall, the links between climate and national security are indirect, involving a complicated web of social and political factors. Climate effects discussed earlier in this report, including food and water security, have the potential to drive national security concerns. Although international cooperation is more typical than conflict in confronting water security issues, conflicts over water issues may become more numerous as droughts become more frequent. In addition, famine and food scarcity have the potential to cause international humanitarian issues and even conflicts, as do health security issues from epidemics and pandemics (also see previous section). These impacts from climate change may present national security challenges through humanitarian crises, disruptive migration events, political instability, and interstate or internal conflict. The impacts on national security are likely to be presented abruptly, in the sense that the eruption of any crisis represents an abrupt change.

An example of an abrupt change that affects the national infrastructure of a number of countries is the opening of shipping lanes in the Arctic as a result of the retreating sea ice. There are geopolitical ramifications related to possible shipping routes and territorial claims, including potential oil, mineral, and fishing rights. The Arctic Council, which was formerly a relatively unknown international body, has become the center of vigorous negotiations over some of these issues. This is a change that is occurring over the course of a couple of decades, well within a generation.

It is important to recognize that abrupt climate change as it affects national security presents opportunities as well as challenges. For example, the United States is still heavily dependent on foreign oil, despite the recent increase in fossil fuel supplies made available by hydrologic fracturing of source rocks. Also, greenhouse gases enter a shared and well-mixed atmosphere, and thus solutions will afford an opportunity to enhance international cooperation and build transparency and trust among nations.

While it may not be possible to predict the exact timing of abrupt climate events and impacts, it is prudent to expect that they will occur at some point. The NRC report on *Climate and Social Stresses* (NRC, 2012b) recommends a scenario approach for preparing for abrupt climate impacts that may have ramifications for national security. The report recommends the use of stress testing, where "a stress test is an exercise to assess the likely effects on particular countries, populations, or systems of potentially disruptive climate events." The material presented in this report could inform these types of stress tests by presenting the types of abrupt climate impacts that are possible (see Chapter 2 and Table 4.1).

CHAPTER 4

The Way Forward

The phrase "Here be dragons" appears on one of the first known globes produced in Europe following Christopher Columbus's voyage across the Atlantic Ocean (Figure 4.1). That globe shows some of the vast uncertainties that Columbus's exploration highlighted for Europe—indeed, North America is portrayed as just a few islands (e.g., De Costa, 1879). The phrase may have been referring to Komodo dragons, but it could easily have evoked the dragons that we now associate with fairy tales in the imagination of Columbus's contemporaries. The initial discovery of the "New World" revealed unsuspected possibilities ("unknown unknowns"), effectively increasing uncertainty. Subsequent exploration has long since filled in those uncertainties, and if real dragons had lived between those islands, they would have been found.

By analogy, many investigations into the climate and Earth system have revealed possible dangers. Some of those have been confirmed or even amplified, such as the impacts of chlorofluorocarbons on the ozone layer that were understood in the 1970s before the Antarctic Ozone Hole was discovered the following decade (see Box 1.1). Other possible dangers, such as the sudden release of methane from ocean sediments, have been greatly reduced by subsequent research that showed a clear lack of dragons.

In looking back at the previous studies of abrupt climate change in the Introduction to this report, the committee notes that even when dragons, i.e., possible threats, are identified and clearly pointed out, they may then be ignored and their presence not acted upon. This is not an unusual situation, and ignoring early warnings is a well-documented phenomenon in environmental research (e.g., EEA, 2001). In this chapter we briefly examine some of the major lessons learned in Chapters 2 and 3 and then propose one possible way forward, namely an Abrupt Change Early Warning System (ACEWS).

WHAT HAS BEEN LEARNED?

Paleoclimatologists have long known that the slower changes of the ice age were punctuated by relatively large events with approximately millennial spacing and sharp onsets and terminations. Some of these were given names, such as the Younger Dryas. A range of studies showed that these events represented changes in wintertime sea

FIGURE 4.1 The Hunt-Lenox Globe. Wording near China says "hic sunt dracones," which translates to "here be dragons." This is a metaphor for unknown threats. Often, as humans have explored more of the world, threats have become less—in this case, the "dragons" may have referred to Komodo dragons, rather than dragons of fairy tales.

ice coverage in the North Atlantic, that they had a near-global footprint but regionally distinct impacts, and that even slow changes in freshwater flux to the North Atlantic could cross a threshold and trigger a sudden event (e.g., Alley, 2007). The very large magnitude of the changes in some regions, the wholesale and rapid reorganization of ecosystems, and the very rapid rates of change that affected certain places (10°C over 10 years in Greenland), together with the realization that greenhouse warming would cause significant changes in freshwater fluxes to the North Atlantic, caused concerns that extended beyond the scientific community into popular culture (e.g., the 2004 science-fiction/disaster movie *The Day After Tomorrow*).

As discussed in Chapter 2, subsequent research has shown that an abrupt disruption of the Atlantic Meridional Overturning Circulation (AMOC) is much less likely under modern boundary conditions than during the ice age, and that the regional cooling impact of the shutdown in heat transport to the high northern latitudes would probably be smaller than the warming from rising greenhouse gas concentrations. In sum, while the North Atlantic remains a probable site of strong climate variability and even slow changes in the circulation of this basin could result in significant local and regional impacts, recent research has shown that the earlier worries of a total AMOC collapse are unwarranted (IPCC, 2007c; USCCSP, 2008; and Chapter 2). Thus in this case, improved knowledge has allayed some fears of the worst types of outcomes occurring this century from this possible abrupt climate change (Box 4.1).

In a similar manner, rapid or catastrophic methane release from sea-floor or permafrost reservoirs has also been shown to be much less worrisome than first considered possible (see section in Chapter 2 on High-Latitude Methane and Carbon Cycles). The discovery of vast methane deposits, clearly vulnerable to warming, motivated discussion of a potential primary role for methane in Earth's climate. Fast changes in atmospheric methane concentration in ice cores from glacial time correlated with abrupt climate changes (e.g., Chappellaz et al., 1993). However, subsequent research has revealed that the variations in methane through the glacial cycles (1) originated in large part from low-latitude wetlands, and were not dominated by high-latitude sources that could be potentially much larger, and (2) produced a relatively small radiative forcing relative to the temperature changes, serving as a small feedback to climate changes rather than a primary driver.[1] Looking to the future, the available source reservoirs for atmospheric methane release—from both methane hydrates and permafrost—are expected to respond to climate on a time scale slow enough that the climate impact from the methane will probably be smaller than that from rising CO_2 concentrations. Nonetheless, there is still much to explore. For example a cause for concern is that wildfires have been spreading into some permafrost regions as local climatic conditions promote increasingly dry conditions (Yoshikawa et al., 2002). Little is known about the potential of such burning to thaw and release stored carbon faster than expected. This possible mechanism of rapid, unexpected carbon release merits research to evaluate its efficacy and climatic impacts.

But despite the comfort one might take from knowing that these examples are not the dragons they were once thought to be, "dragons" in the climate system still may exist.

[1] Methane was also proposed as the origin of the Paleocene–Eocene thermal maximum event, 55 million years ago, in which carbon isotopic compositions of $CaCO_3$ shells in deep sea sediments reflect the release of some isotopically light carbon source (like methane or organic carbon), and various temperature proxies indicate warming of the deep ocean and hence the Earth's surface. But the longevity of the warm period has shown that CO_2 was the dominant active greenhouse gas, even if methane was one of the important sources of this CO_2, and the carbon isotope spike shows that if the primary release reservoir were methane, the amount of CO_2 that would be produced by this spike would be insufficient to explain the extent of warming, unless the climate sensitivity of Earth was much higher than it is today (Pagani et al., 2006).

BOX 4.1 ASYMMETRICAL UNCERTAINTIES

The current understanding of many aspects of the climate system and its influence on ecosystems and economies projects a most-likely response to changing CO_2 but a skewed distribution of uncertainties, such that the outcome may be a little "better" (smaller changes, lower costs), a little "worse" (larger changes, higher costs), or a lot worse than the most-likely case, but with little chance of being a lot better (see Figure). Skewed distributions, and those with "fat tails" that allow finite chances of very large changes, are known or have been suggested in many contexts (e.g., Mandelbrot, 1963; Weitzman, 2011). Vigorous research continues on the effects of such issues on optimal policy paths (e.g., Keller et al., 2004; Keller et al., 2008; McInerney and Keller, 2008; Weitzman, 2009; Tol, 2003). As described above, research can in some cases lead to an improved understanding of risks where the highest impact outcomes are discovered to be less likely than originally feared, i.e., in some cases research can remove these "fat tails" (see Figure).

The value of reducing the deep uncertainty associated with fat tails or abrupt climate changes is clear, motivating research to stimulate learning. A large possibility of an AMOC "shutdown", for example, would have notably increased the optimal response to reducing climate change, so the reduction of the estimated chance of such an event provided by recent research has large economic consequences (Keller et al., 2004).

Many of the potential abrupt climate changes listed in Table 4.1 are listed as having a "moderate" or "low" probability/likelihood of occurring within this century. Recent research on expert elicitations (Kriegler et al., 2009) related to the probability of abrupt climate changes is in general agreement with the assessments provided in Table 4.1. For the five tipping point examples that they examined, they found "significant lower probability bounds for triggering major changes in the climate system."

The collected understanding of these threats is summarized in Table 4.1. For example, the West Antarctic Ice Sheet (WAIS) is a known unknown, with at least some potential to shed ice at a rate that would in turn raise sea level at a pace that is several times faster than is happening today. If WAIS were to rapidly disintegrate, it would challenge adaptation plans, impact investments into coastal infrastructure, and make rising sea level a much larger problem than it already is now. Other unknowns include the rapid loss of Arctic sea ice and the potential impacts on Northern Hemisphere weather and climate that could potentially come from that shift in the global balance of energy, the widespread extinction of species in marine and terrestrial systems, and the increase in the frequency and intensity of extreme precipitation events and heat waves. The committee reviews the various abrupt changes described in Chapter 2 in Table 4.1.

Looking beyond the physical climate system, many of the research frontiers highlighted in this report focus on the potential for gradual physical changes to trigger

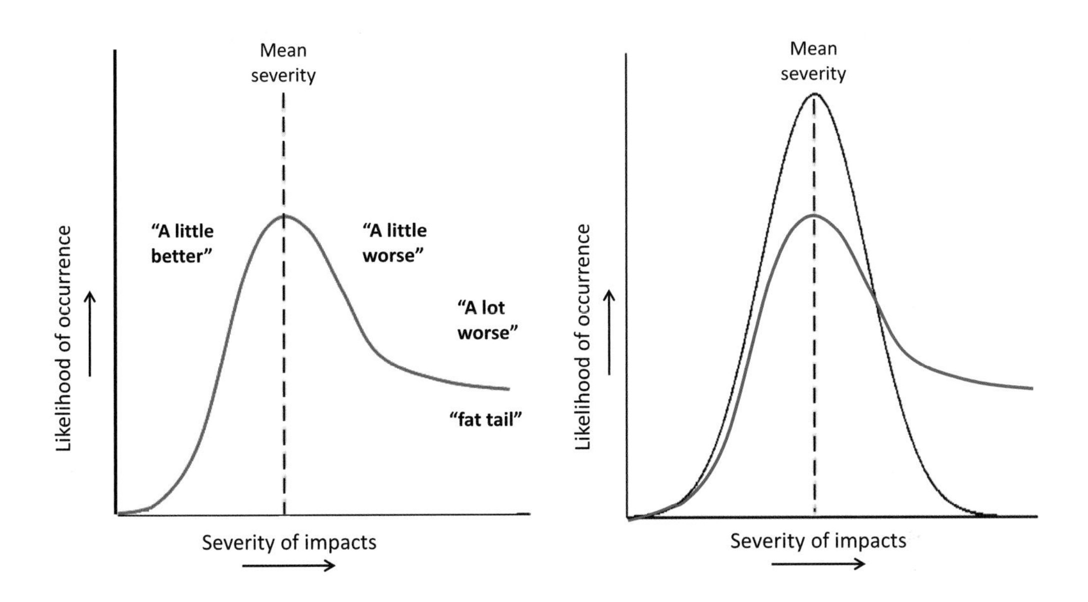

FIGURE The graph on the left represents the skewed distribution of uncertainties with a "fat tail." The mean likelihood of occurrence at the level of severity anticipated is represented by a dotted line. The area to the left of the mean represents the likelihood for impacts less severe (the "a little better" case), while the area to the right shows the greater likelihood for extreme impacts (spanning "a little worse" to "a lot worse" cases). The graph on the right compares the normal distribution (black line) to the "fat tail" distribution (blue line). For some changes, more research has shown that the distribution of possible outcomes includes less likelihood of the most severe outcomes.

abrupt ecological, economic, or social changes (see Chapter 3). In these cases, reaching a certain threshold of change in various climatic parameters can trigger abrupt, irreversible changes in the affected ecological, economic, or social system, even if the trajectory of climate change itself is gradual. There is still much to learn about the potential for and possible prediction of these kinds of abrupt changes, but a sound body of theory and empirical data (Barnosky et al., 2012; Carpenter et al., 2011; Hastings and Wysham, 2010; Mumby et al., 2007; Scheffer, 2010; Scheffer et al., 2001, 2009, 2012b) confirm that there are real "dragons" out there to be discovered.

In the ecological arena, some of these metaphorical dragons are already becoming evident—for instance, gradual ocean acidification leading to the shutdown of developmental pathways in ecologically important marine species, or gradual warming of ocean waters to exceed temperature thresholds that result in death of coral reefs (see sections on Changes in Ocean Chemistry and Extinctions in Chapter 2). Both kinds of "threshold effects" have been confirmed experimentally (ocean acidification) or from observational data accumulated over the past few decades (coral bleaching and death).

A major area of uncertainty, however, is the potential for threshold-induced impacts on one species to cascade to impact others, and the pathways by which such cascades could lead to wholesale ecosystem collapse (see section on Ecosystem Collapse and Rapid State Changes in Chapter 2). For example, while it is clear that species are changing their geographic distributions and seasonal cycles individualistically in response to climate change, resulting in pulling apart of species that have co-existed in the same place at the same time of the year, the magnitude and importance of longer-term ecological changes that will result are still unclear. Answering such species-interaction and ecological-network questions holds the key to assessing the likelihood of figuring out the 'fat-tail' probabilities of ecosystem collapse induced by climate change, which in turn would impact economic and social systems (for instance, through loss of fisheries, forests, or agricultural productivity).

Anticipating the potential for climatically-induced abrupt change in social systems is even more difficult, given that social systems are actually extremely complex systems, the dynamics of which are governed by a network of interactions between people, technology, the environment, and climate. The sheer complexity of such systems makes it difficult to predict how changes in any single part of the network will affect the overall system, but theory indicates that changes in highly-connected nodes of the system have the most potential to propagate and cause abrupt downstream changes. Climate connects to social stability through a wide variety of nodes, including availability of food and water, transportation (for instance, opening Arctic seaways), economics (insurance costs related to extreme weather events or rising sea level, agricultural markets, energy production), ecosystem services (pollination, fisheries), and human health (spread of disease vectors, increasing frequency of abnormally hot days that cause physiological stress). Reaching a climatic threshold that causes rapid change in any one of these arenas therefore has high potential to trigger rapid changes throughout the system. For example, at the time Arctic shipping routes become routinely passable, world trade routes and the related economic and political realities could change dramatically within a single decade.

Much remains unknown about how climate maps onto the complex networks that define social systems at local, regional, and global scales. Using network modeling techniques to identify the nodes and connections that construct social systems at a variety of spatial scales, and how projected climate changes would be expected to propagate through the system, may well lead to better predictive ability. A more empirical approach, already feasible with existing modeling techniques, is to identify communities that are geographically situated where small shifts in global climatic patterns—such as in position of the jet stream—would have large impacts at the local scale, such as an "on-off" switch for local drought. Identifying those geographic areas and assessing how the local impacts might propagate spatially through the regional and global social network may provide a viable means of anticipating regional vulnerabilities, and which of those vulnerable regions have high potential of influencing global dynamics. "Stress tests," or scenario based modeling exercises have been recommended as a way to reveal vulnerabilities and likely effects of disruptive climate events on particular countries, populations, or systems; stress tests provide "a framework for integrating climate and social variables more systematically and consistently within national security analysis" (NRC, 2012b).

ANTICIPATING SURPRISES

The recognition of the importance of tipping-point behavior in physical, biological, and social systems has prompted a growing body of research to provide as much early warning as possible of incipient or ongoing abrupt changes (Box 4.2). Theory and experiment agree that some systems approaching tipping points exhibit signs of the impending change. This behavior may include a flickering behavior, in which a system jumps back and forth between two states, or a shift to slower recovery from small perturbations (e.g., Taylor et al., 1993; Drake and Griffen, 2010; Carpenter et al., 2011; Veraart et al., 2012; Wang et al., 2012a), presaging future failure to recover. However, as emphasized by Boettinger and Hastings (2013), there probably is not a generic signal of an impending shift, with different signals in different systems, and there is very real possibility that no warning signal will be evident. Considering this, plus the various challenges to interpreting signals that do occur, the goal of successfully predicting tipping points, and providing policy-relevant choices on how to avoid them or deal with the consequences, is likely to be realized in some cases but is unlikely to be universally possible.

Nonetheless, identifying potential vulnerabilities is valuable. In some cases science may be able to provide accurate information that a tipping point is imminent, creating time for adaptation or even possibly mitigation. Science can also identify when tipping is occurring, or has recently occurred. Such knowledge could greatly reduce societal damages. And, in light of the potentially very large costs of some possible abrupt changes, additional research to improve this knowledge can have large economic benefits (e.g., Keller et al., 2007; McInerney et al., 2012).

BOX 4.2 EARLY WARNING SIGNALS OF ABRUPT TRANSITIONS

Examining the spectral properties of time series data can, in theory, predict the onset of some abrupt changes. There is a burgeoning literature examining early warning signals in systems as they approach tipping points (e.g., Boettiger and Hastings, 2012a, b; Dakos et al., 2008; Ditlevsen and Johnsen, 2010; Lenton et al., 2008; Scheffer et al., 2009; Scheffer et al., 2012a). Several examples of early warning signs are described below:

1. In some systems, the return time from a small perturbation increases as the system approaches a critical threshold. This "critical slowing down" leads to an increase in autocorrelation in the pattern of variability, which can serve as an indicator of impending abrupt change (Dakos et al., 2008; Scheffer et al., 2009).
2. There may also be "flickering" as stochastic forcings move the system back and forth across a threshold to sample two alternative regimes; some data suggest that past climatic shifts may have been preceded by such flickering behavior (Scheffer et al., 2009).
3. It is also possible for the macrostructure of systems to indicate proximity to a transition point (e.g., spatial patterns of vegetation in a landscape as it transitions from patchy to barren; Scheffer et al., 2009). Research in this field is extending to examine highly connected networks, where connectivity and heterogeneity patterns may be used to anticipate state changes (Scheffer et al., 2012a).

Significant challenges exist in implementing early warnings to anticipate tipping points. Accurate forewarnings that avoid significant false alarms involve tradeoffs between specificity and sensitivity (Boettiger and Hastings, 2012a, b). Additionally, successful detection of an impending change does not imply that an effective intervention is possible. Finally, there is some evidence that previous abrupt changes in Earth's history have been noise-induced transitions (see Box 1.4); such events will have very limited predictability (Ditlevsen and Johnsen, 2010). In general, because many systems have the property of being sensitive to initial conditions (i.e., the "butterfly effect"), forecasting future behavior is only feasible on short timescales for those systems, and timely early warning signs for abrupt changes may not be possible.

An Abrupt-Change Early-Warning System

In light of the potential great importance and value of accurate anticipation of the occurrence and impacts of abrupt changes, the committee recommends development of an abrupt change early warning system (ACEWS). An ACEWS would be part of an overall risk management strategy, providing required information for hazard identification and risk assessment (Figure 4.2). This information would then inform an overall risk strategy, which uses the risk assessment to prioritize hazard mitigation options and their implementation (Basher, 2006). A central part of an effective risk management framework is continued evaluation of the efficacy of the risk management strat-

FIGURE 4.2 Continuous and reinforcing process of disaster risk management as a foundation for building resilience. Central to the risk management process is the collective evaluation by the partners regarding goals, values, and objectives for the risk management strategy and for community resilience. The entire process, divided for convenience of discussion into six steps, encompasses the ability to identify and assess the local hazards and risks (steps 1 and 2), to make decisions as to which strategies or plans are most effective to address those hazards and risks and implement them (steps 3 and 4), and to review and evaluate the risk management plan and relevant risk policies (steps 5 and 6). An Abrupt Change Early Warning System (ACEWS) would be part of such an overall risk management strategy, providing required information for hazard identification and risk assessment (adapted from NRC, 2012c).

egy and adjustment based on new information about the hazard and vulnerabilities (NRC, 2012c). Given the number of critical Earth system components that might cross tipping points and the potentially data intensive monitoring and modeling needs, some strategic decisions will have to be made with regard to which hazards to monitor and for which to attempt to develop warning systems. For example, if the lead time necessary to prepare the appropriate social response is very long, the detection of the tipping point might not occur early enough to avoid major impact on the socioeconomic system. Some of these decisions are based on what risks are socially acceptable compared to the cost of the hazard mitigation effort and involve value judgment by the affected people or their political representatives (Plattner, 2005).

In general, an ACEWS system would (1) identify and quantify social and natural vulnerabilities and ensure long-term, stable observations of key environmental and economic parameters through enhanced and targeted monitoring, (2) integrate new knowledge into numerical models for enhanced understanding and predictive capability, and (3) synthesize new learning and advance the understanding of the Earth system, taking advantage of collaborations and new analysis tools. These aspects are discussed below, followed by a discussion of the some special considerations for designing and implementing an ACEWS.

The development of an ACEWS will need to be an ongoing process, one that goes beyond the scope of this report to include multiple stakeholders. As such, there are numerous nuances and issues not addressed here. Vulnerabilities will need to be prioritized, and how the needs and desires of various stakeholder groups are considered can change relative priorities. Some economic costs are clear (threats to an airport from sea level rise, for example) and some are less clear (potential loss of ecosystem services, for example), making triage difficult. It is noted that communication is a crucial component of any early warning system to ensure the timely delivery of information on impending events, and prepare potential risk scenarios and preparedness strategies. Special considerations need to be given to the importance of accuracy, lead time, warning message content, warning transmission, and the appropriate social response to minimize negative consequences from the hazard (Kasperson et al., 1988; Mileti, 1999). For example, the fact that Superstorm Sandy was not labeled a hurricane by the National Hurricane Center as it came on shore (the storm lost its tropical characteristics before landfall) may have confused those in harm's way and led unnecessarily to a lessened sense of urgency and danger.[2] In addition, an overall risk management system requires a preparedness and adaptations sub-system that feeds back on

[2] For example, http://www.wsfa.com/story/21807734/whats-in-a-name-sandy-hurricane-or-superstorm.

BOX 4.3 LESSONS LEARNED FROM EARLY WARNINGS IN PAST ENVIRONMENTAL ISSUES

Lessons from the European Environment Agency 2001 report on "Late lessons from early warnings: the precautionary principle 1896-2000."

1. Acknowledge and respond to ignorance, as well as uncertainty and risk, in technology appraisal and public policymaking.
2. Provide adequate long-term environmental and health monitoring and research into early warnings.
3. Identify and work to reduce "blind spots" and gaps in scientific knowledge.
4. Identify and reduce interdisciplinary obstacles to learning.
5. Ensure that real world conditions are adequately accounted for in regulatory appraisal.
6. Systematically scrutinize the claimed justifications and benefits alongside the potential risks.
7. Evaluate a range of alternative options for meeting needs alongside the option under appraisal, and promote more robust, diverse and adaptable technologies so as to minimize the costs of surprises and maximize the benefits of innovation.
8. Ensure use of "lay" and local knowledge, as well as relevant specialist expertise in the appraisal.
9. Take full account of the assumptions and values of different social groups.
10. Maintain the regulatory independence of interested parties while retaining an inclusive approach to information and opinion gathering.
11. Identify and reduce institutional obstacles to learning and action.
12. Avoid "paralysis by analysis" by acting to reduce potential harm when there are reasonable grounds for concern.

SOURCE: EEA, 2013.

loss and damage by informing actions needed to reduce impacts from an impending event.

An excellent summary of lessons learned from early warnings in past environmental issues can be found in a 2001 report by the European Environment Agency, and are shown in Box 4.3. In this section, the Committee provides further thoughts on selected aspects of an ACEWS: the monitoring, modeling, and synthesis aspects, as well as some special considerations for designing and implementing an ACEWS.

Monitoring

An ACEWS will require sustaining and integrating existing observing capabilities, as well as adding new capabilities targeted at improving understanding or early warnings relevant to specific abrupt change threats. Table 4.1 provides initial thoughts on the monitoring needs for each of the abrupt changes considered in this report. In general, observations involved in an ACEWS will include high resolution paleo-climate records that can adequately resolve and sample extreme events, field experiments that help improve our understanding of the thresholds and tipping points of various elements of Earth's climate system, and carefully calibrated long-term, global ground-based, airborne, and satellite observing systems. Because the most vulnerable regions tend to be resource limited or inaccessible for reliable ground-based observations, satellite observations are critical.

For the purpose of anticipating climate-related surprises through an ACEWS, thoughtful consideration is needed of what must be observed, on what timescales, and with what accuracy. The detailed discussions in this report (as summarized in Table 4.1) provide initial information on these needs. These are based on previous research on what aspects of the system may undergo abrupt change and what mechanisms lead to or contribute to that change. To further refine and inform monitoring needs will require additional research on relevant processes. This is particularly true when looking at the integrated system as a whole and the interconnections across the physical, biogeochemical, and societal spheres. The vision of an ACEWS is to build up from monitoring systems that look at individual processes to eventually develop an integrated and coordinated monitoring system. Such coordination is needed to facilitate the understanding of the interconnectedness of the various individual processes and avoid the compartmentalized thinking that can lead to duplication of effort as well as missing key processes that occur at the interfaces between disciplines.

As a general approach for examining what monitoring needs are required for a specific potential abrupt change, one[3] could consider the decision tree in Box 4.4.

Modeling

Another key part of an ACEWS is modeling. Numerical models, from reduced-complexity dynamical systems (Eisenman and Wettlaufer, 2009) to complex global coupled Earth system models (Holland et al., 2006), have provided unique insights into the mechanisms and likelihood for abrupt climate change (Hodell et al., 1995; Alley and Joughin, 2012). A successful and adaptive ACEWS must consistently iterate between data collection, model testing and improvement, and model predictions that suggest better data collection. A successful example of this model-monitoring nexus

[3] Who might be making these decisions is discussed in the implementation section below.

is the RAPID array, in which the monitoring strategy was initially tested in numerical models, which in turn informed improved modeling strategies.[4]

As discussed in Chapters 2 and 3, many of the processes involved in abrupt impacts occur at relatively small spatial scales that are not well resolved in current models. Strategies to address this shortcoming need to be proposed, vetted, and employed by climate modelers and stakeholders. Related to this, the implications of model biases and simplifications for the simulation of abrupt change are in general not well established, although there are some examples in which a thorough analysis on this has been performed (e.g., Eisenman, 2012). Further studies on the role of model biases and simplifications on feedbacks of interest are necessary to inform model developments and relevant applications. In addition, large ensemble simulations are important for adequately characterizing the probability of extreme events in the past, present, and future, as well as to support developments of scenarios for societal stress tests and plans for observing systems.

To date, modeling studies have had less focus on how gradual changes in climate could induce abrupt changes in other aspects of the system. This is in large part because many of these models have not incorporated relevant processes in their formulation. Full Earth system models (beyond just atmosphere and ocean models) have only recently been developed, and work is underway to improve the Earth system-related components of these models, such as active carbon cycles, atmospheric chemistry, and ice sheets. Inclusion of these model components will enable studies on many of the important feedbacks in the Earth system, particularly those positive feedbacks that amplify change and can send systems past tipping points that in part characterize abrupt change.

Synthesis

Monitoring and modeling needs are critical aspects of establishing an effective ACEWS. However, they cannot exist alone. A successful ACEWS must avoid the trap of data collection without continuing and evolving data analysis and model integration. A necessary third part of the proposed ACEWS is synthesizing knowledge. The ACEWS will require:

- Dedicated people who will apply their expertise in collaborative activities to assess the potential for abrupt change in a broader systems perspective;

[4] RAPID integrates multiple sensors with international support, and RAPID integrates science in the monitoring, to extract maximum value from the effort, and to avoid loss of key insights in unexamined data.

BOX 4.4 A DECISION TREE FOR EXAMINING MONITORING NEEDS

As a first step, identify the known and suspected thresholds. Some triage needs to be employed initially to avoid the temptation to monitor everything, and to ensure that the most obvious threats are prioritized. Prior reports and this one provide guidance, although additional effort may be needed in social and economic areas.

As a second step, ask whether there is there an existing monitoring system already in place that could be used as is, or modified to meet the needs of an ACEWS (see Figure). In many cases, science already monitors key functions of the climate and other Earth cycles, and these systems should be exploited in an ACEWS. They may need more frequent data updates, or data handling may need to be modified to better use the data.

If the answer to the above question is yes, then an important action is to protect that network to ensure that it continues to operate. A security camera that does not work, or does not watch all entrances and exits, is not very useful. Current resources to provide high-quality and continuous monitoring are at risk, and there are notable examples including critical time series that have been compromised due to reductions of in-situ and remotely sensed observational networks (e.g., the NOAA Cooperative Air Sampling Network). To improve our understanding of the evolving Earth system, monitoring resources must be protected and in some cases expanded. Another example of this category of monitoring system is the GRACE satellite mission. Originally launched in 2002, GRACE (actually two satellites) measures gravity, which in continental areas varies largely as a function of overall water content, and thus highlights areas with changing

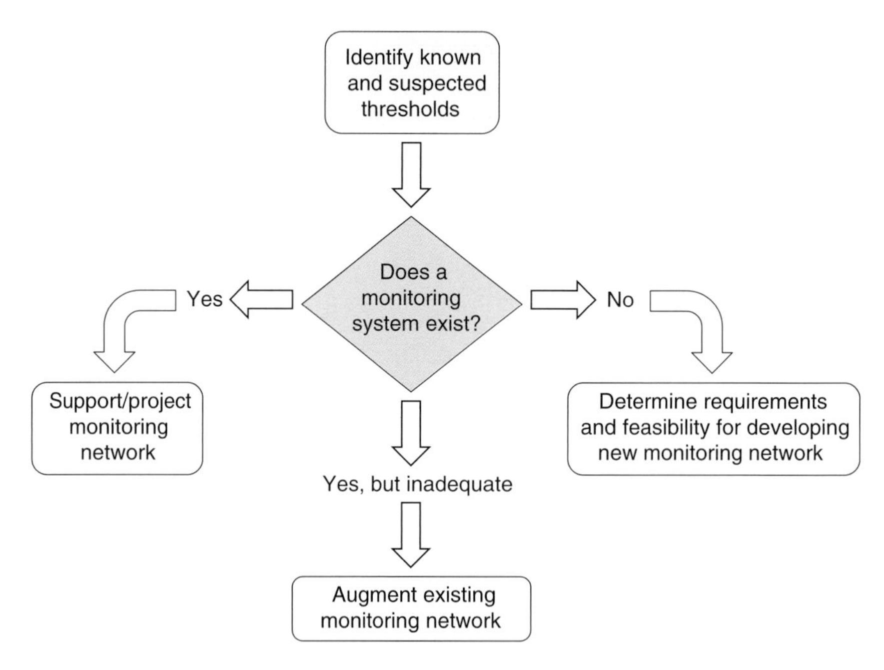

FIGURE Simplified ACEWS decision tree for monitoring.

BOX 4.4 CONTINUED

groundwater storage or withdrawal as well as changes in glaciers and ice sheets. In its 10-year lifetime, GRACE has revealed unsustainable rates of groundwater withdrawal in the southeastern United States and in parts of India, for example, and obtaining such data from other sources would be quite difficult in the United States and much more so in India. A successful ACEWS that monitors for food security should include such monitoring, as groundwater withdrawal is the common approach to combatting the impact of drought on crop yields. A failure of groundwater to backstop rain would be a tipping point in the production of food, and a central part of a system to forecast famine.[a]

If the answer to the above question is no (there is no existing monitoring system in place), then the next steps are to determine what is needed to implement a monitoring system, and to do so if it is feasible. A national inventory of the types and values of coastal resources vulnerable to some set level of sea level rise and storm surge is an example of such a monitoring system, one that does not currently exist but could be created and updated using information that is currently collected, but not collated into a central database.[b] Another example is a system to monitor the interaction between the ocean and the outlet glaciers of land ice. It is clear that ocean currents and temperatures play a key role in melting ice (Joughin et al., 2012a), and will be important in monitoring for catastrophic instability in the marine-based West Antarctic Ice Sheet as well as marine portions of the Greenland and East Antarctic ice sheets (Chapter 2, section on Ice Sheets and Sea Level), which in turn present the greatest risk of an unexpected and rapid contribution to sea level rise.

A third possible answer to the above question is yes, a current monitoring system exists but it is inadequate to meet the needs of an ACEWS and it therefore needs to be augmented. An example of this is the Circumpolar Active Layer Monitoring (CALM) program, which is designed to observe the response of near-surface permafrost to climate change over long (multi-decadal) time scales. CALM, which is part of the National Science Foundation's Arctic Observing Network, would likely need to be expanded and more automated to function as an ACEWS, especially because, as described in the section on Potential Climate Surprises Due to High-Latitude Methane and Carbon Cycles in Chapter 2, there are large geographic coverage gaps in important regions.

The science of abrupt climate change is not settled; monitoring needs will evolve over time and an iterative mechanism would allow ongoing assessment and evaluation. There needs to be some mechanism to allow for evolution of the ACEWS. One way to do this would be the creation of an ACEWS steering committee that would regularly visit the state of ACEWS monitoring efforts, critically examine proposals for new monitoring systems, and ensure that the current systems are meeting their stated goals. This is described more below as part of the implementation section.

[a] Note: A GRACE follow-on mission is currently planned (see the FY 2014 President's budget: http://www.nasa.gov/pdf/740512main_FY2014%20CJ%20for%20Online.pdf).

[b] Any inventory effort could build upon existing efforts like those that follow on from work of the Environmental Protection Agency (http://plan.risingsea.net/index.html) and Climate Central (http://sealevel.climatecentral.org/).

- Improved collaborative networks to entrain new communities of scientists and researchers to take fresh looks at the problem; for example, integrating climate scientists with applied mathematicians, statisticians, dynamical systems experts, policy analysts, engineers, city planners, ethicists and others who ensure that the ACEWS meets the needs of a broad spectrum of stakeholders will result in a better ACEWS that produces products that are more useful to those affected;
- Enhanced educational activities to provide a platform for innovation in producing a workforce that is comfortable working across the boundaries; and
- Innovative tools—including new data analysis and modeling techniques, to allow for a novel perspective on abrupt change, as well as more robust statistical tools that are needed for analyzing and understanding non-linear dynamic systems and inter-connection among various climate fields; in some cases, this will include applications of tools created in a different context to the abrupt change problems.

These research elements are a key part of any monitoring or early warning system efforts.

Special Considerations for Designing and Implementing an ACEWS

Although the committee discusses some general concepts about implementing an ACEWS below, the implementation of an ACEWS should be planned and executed by those agencies that are tasked to do so and/or contribute funding to the effort. How that effort is organized is beyond the purview of the committee, and thus the ideas below should be viewed as suggestions intended to be helpful, and not prescriptive.

One important aspect of an ACEWS should be the integration of the effort, from the monitoring, through modeling and interpreting of the data, to producing scientific products including peer-reviewed publications and consumer-friendly data products. Examples of monitoring efforts in which the data have not been analyzed regularly are unfortunately common,[5] and given the inherently time-sensitive nature of an ACEWS, such an outcome would be anathema to the intent of the system. A few examples of current monitoring programs that integrate monitoring with active interpretation of the data include the Long Term Ecological Programs of the National Science Foundation, NOAA's atmospheric gas monitoring program, NASA's Stratospheric Observatory

[5] The initial failure of satellite monitoring to identify the ozone hole over Antarctica because an automatic routine flagged anomalous data as potential errors shows what can happen without sufficiently integrated interpretation (see Grundmann, 2002 for a more complete historical account).

(SOFIA), and the National Ecological Observatory Network (NEON). Those programs that are currently operational fully integrate an active science program with observations, and have all been extremely successful in their respective missions. An ACEWS should build upon the success of these programs and others like them.

Being mindful of stakeholder priorities and meeting stakeholder needs is another aspect of a successful ACEWS that should be stressed. The National Integrated Drought Information System (NIDIS) is an example of a successful early warning program that integrates the needs of the user community with the monitoring and modeling systems. NIDIS incorporates a spectrum of drought-related products, including regional products tailored to the needs of specific regions, as a regular part of the system. In general, research focus and early warning signal detection would be most beneficial if they were prioritized based on societal impacts and likelihood of occurrence of the extreme events and resultant abrupt changes.

Another important aspect of a successful ACEWS is for the system to be flexible and adaptive. It is not enough to integrate data and interpretation, but the science involved in the interpretation should inform the whole system and help it to evolve to better meet the needs of society. This step is important for an ACEWS as in some cases we know what to watch for, such as changes in AMOC, and in some cases we are less sure, such as changes in Northern Hemisphere weather patterns that may accompany the large energy changes in the Arctic as sea ice melts, trading white sea ice that reflects solar energy for blue ocean that absorbs solar energy. Also, the system should be nimble enough to change focus if necessary as knowledge about abrupt change improves. It is clear today that there is much to learn about the threats of abrupt change, and an ACEWS would be best served if it were designed to evolve as knowledge, monitoring tools, and societal needs, evolve. NIDIS is an example of a system that strives to meet this goal. For example, the NIDIS Regional Drought Early Warning Systems explore a variety of early warning and drought risk reduction strategies, seeking to match user needs with observations and analyses while allowing for system adaptation and evolution.

Organization of an ACEWS would benefit from capitalizing on existing programs, but there will be a need to capture the interconnectedness of the various parts of the climate and human systems. ACEWS could eventually be run as a large, overarching program, but might better be started through coordination, integration, and expansion of existing and planned smaller programs like the Famine Early Warning System Network (FEWS NET), which is "a USAID-funded activity that collaborates with international, regional, and national partners to provide timely and rigorous early warning and vulnerability information on emerging and evolving food security issues."[6] One possible mechanism to eventually achieve this overall coordination would be to start with a steering group who could provide efficient guidance. Such a steering committee could be made up of representatives of funding agencies, scientists, representatives of various user communities (including national security and interested businesses), and international partners, to name a subset of the possibilities. Subgroups or working groups may be able to bring focus to specific issues that require more attention as needed, e.g., water, food, or ecosystem services. Beyond a steering group, a number of interagency coordinating mechanisms exist, and the committee is not specifically recommending one over another. Whatever the mechanism, the committee does stress that coordination—to reduce duplication of efforts, maximize resources, and facilitate data and information sharing—is key to a successful ACEWS.

ACEWS: NEED FOR ACTION

As noted earlier, the proper design and implementation of an ACEWS will need to be an ongoing process and will require expertise from many different disciplines beyond just the physical sciences, as well as input from many different stakeholder groups. Providing a complete roadmap to a successful ACEWS was beyond the scope of this report, but the committee has outlined its initial thoughts on what would make such a system successful above. Much is known about the design, implementation, and sustainability of early warning systems that can be leveraged in addition what is described in this report. In summary, this report should be viewed as a call for an ACEWS to be designed, for to not make such a call would be to willfully ignore "dragons," and that was an approach the committee strongly opposed.

The committee views this call as being particularly salient in light of its analysis of the previous reports on abrupt climate change, where a common theme emerged. Beginning with the 2002 NAS study (NRC, 2002), recommendations have been made to address the problem, but little follow-up action has been taken. To gain the benefits that science can offer on this topic, action is needed now. An ACEWS need not be overly expensive and need not be created from scratch, as many resources now exist that can contribute, but the time is here to be serious about the threat of tipping points so as to better anticipate and better prepare ourselves for the inevitable surprises.

[6] http://www.fews.net/Pages/default.aspx.

"No matter how clear our foresight, no matter how accurate our computer models, a belief about the future should never be mistaken for the truth.

The future, as such, never occurs. It becomes the present.

And no matter how well we prepare ourselves, when the imagined future becomes the very real present, it never fails to surprise."

—Alan AtKisson, Believing Cassandra

TABLE 4.1 State of knowledge on potential candidate processes that might undergo abrupt change. These include both abrupt climate changes in the physical climate system and abrupt climate impacts of ongoing changes that, when certain thresholds are crossed, can cause abrupt impacts for society and ecosystems. The near term outlook for this century is highlighted as being of particular relevance for decision makers generally.

	Potential Abrupt Climate Change or Impact and Key Examples of Consequences	Current Trend	Near Term Outlook (for an Abrupt Change within This Century)	Long Term Outlook (for a Significant Change[1] after 2100)	Level of Scientific Understanding	Critical Needs (Research, Monitoring, etc.)
Abrupt Changes in the Ocean	**Disruption to Atlantic Meridional Overturning Circulation (AMOC)** • Up to 80 cm sea level rise in North Atlantic • Southward shift of tropical rain belts • Large disruptions to local marine ecosystems • Ocean and atmospheric temperature and circulation changes • Changes in ocean's ability to store heat and carbon	Trend not clearly detected	**Low**	High	Moderate	• Enhanced understanding of changes at high latitudes in the North Atlantic (e.g., warming and/or freshening of surface waters) • Monitoring of overturning at other latitudes • Enhanced understanding of drivers of AMOC variability
Abrupt Changes in the Ocean	**Sea level rise (SLR) from ocean thermal expansion** • Coastal inundation • Storm surges more likely to cause severe impacts	Moderate increase in sea level rise	**Low[2]**	High	High	• Maintenance and expansion of monitoring of sea level (tide gauges and satellite data), ocean temperature at depth, local coastal motions, and dynamic effects on sea level
Abrupt Changes in the Ocean	**Sea level rise from destabilization of WAIS ice sheets** • 3-4 m of potential sea level rise • Coastal inundation • Storm surges more likely to cause severe impacts	Losing ice to raise sea level	**Unknown but Probably Low**	Unknown	Low	• Extensive needs, including broad field, remote-sensing, and modeling research
Abrupt Changes in the Ocean	**Sea level rise from other ice sheets (including Greenland and all others, but not including WAIS loss)** • As much as 60m of potential sea level rise from all ice sheets • Coastal inundation • Storm surges more likely to cause severe impacts	Losing ice to raise sea level	**Low**	High	High for some aspects, Low for others	• Maintenance and expansion of satellite, airborne, and surface monitoring capacity, process studies, and modeling research

TABLE 4.1 Continued

	Potential Abrupt Climate Change or Impact and Key Examples of Consequences	Current Trend	Near Term Outlook (for an Abrupt Change within This Century)	Long Term Outlook (for a Significant Change[1] after 2100)	Level of Scientific Understanding	Critical Needs (Research, Monitoring, etc.)
...in the Ocean (cont.)	**Decrease in ocean oxygen (expansion in oxygen minimum zones [OMZs])** • Threats to aerobic marine life • Release of nitrous oxide gas—a potent greenhouse gas—to the atmosphere	Trend not clearly detected	**Moderate**	High	Low to Moderate	• Expanded and standardized monitoring of ocean oxygen content, pH, and temperature • Improved understanding and modeling of ocean mixing • Improved understanding of microbial processes in OMZs
Abrupt Changes in the Atmosphere	**Changes to patterns of climate variability (e.g., ENSO, annular modes)** • Substantial surface weather changes throughout much of extratropics if the extratropical jetstreams were to shift abruptly	Trends not detectable for most patterns of climate variability Exception is southern annular mode—detectable poleward shift of middle latitude jetstream	**Low**	Moderate	Low to Moderate	• Maintaining continuous records of atmospheric pressure and temperatures from both in-situ and remotely sensed sources • Assessing robustness of circulation shifts in individual ensemble members in climate change simulations • Developing theory on circulation response to anthropogenic forcing
Abrupt Changes in the Atmosphere	**Increase in intensity, frequency, and duration of heat waves** • Increased mortality • Decreased labor capacity • Threats to food and water security	Detectable increasing trends	**Moderate** (Regionally variable, dependent on soil moisture)	High	High	• Continued progress on understanding climate dynamics • Increased focus on risk assessment and resilience
Abrupt Changes in the Atmosphere	**Increase in frequency and intensity of extreme precipitation events (droughts/floods/hurricanes/major storms)** • Mortality risks • Infrastructure damage • Threats to food and water security • Potential for increased conflict	Increasing trends for floods Trends for drought and hurricanes not clear	**Moderate**	Moderate to High	Low to Moderate	• Continued progress on understanding climate dynamics • Increased focus on risk assessment and resilience

TABLE 4.1 Continued

	Potential Abrupt Climate Change or Impact and Key Examples of Consequences	Current Trend	Near Term Outlook (for an Abrupt Change within This Century)	Long Term Outlook (for a Significant Change[1] after 2100)	Level of Scientific Understanding	Critical Needs (Research, Monitoring, etc.)
Abrupt Changes at High Latitudes	**Increasing release of carbon stored in soils and permafrost** • Amplification of human-induced climate change[3]	Neutral trend to small trend in increasing soil carbon release	**Low**	High	Moderate[4]	• Improved models of hydrology/cryosphere interaction and ecosystem response • Greater study of role of fires in rapid carbon release • Expanded borehole temperature monitoring networks • Enhanced satellite and ground-based monitoring of atmospheric methane concentrations at high latitudes
	Increasing release of methane from ocean methane hydrates • Amplification of human-induced climate change	Trend not clearly detected	**Low[5]**	Moderate	Moderate[6]	• Field and model based characterization of the sediment column • Enhanced satellite and ground-based monitoring of atmospheric methane concentrations at high latitudes
	Late-summer Arctic sea ice disappearance • Large and irreversible effects on various components of the Arctic ecosystem • Impacts on human society and economic development in coastal polar regions • Implications for Arctic shipping and resource extraction • Potential to alter large-scale atmospheric circulation and its variability	Strong trend in decreasing sea ice cover	**High**	Very high	High	• Enhanced Arctic observations, including atmosphere, sea ice, and ocean characteristics • Better monitoring and census studies of marine ecosystems • Improved large-scale models that incorporate the evolving state of knowledge
	Winter Arctic sea ice disappearance • Same as late summer Arctic sea ice disappearance above, but more pronounced due to year-round lack of sea ice	Small trend (Decreasing but not disappearing)	**Low**	Moderate	High	• Same as late summer Arctic sea ice disappearance above

TABLE 4.1 Continued

	Potential Abrupt Climate Change or Impact and Key Examples of Consequences	Current Trend	Near Term Outlook (for an Abrupt Change within This Century)	Long Term Outlook (for a Significant Change[1] after 2100)	Level of Scientific Understanding	Critical Needs (Research, Monitoring, etc.)
Abrupt Changes in Ecosystems	**Rapid state changes in ecosystems, species range shifts, and species boundary changes** • Extensive habitat loss • Loss of ecosystem services • Threats to food and water supplies	Species range shifts significant; others not clearly detected	**Moderate**	High	Moderate	• Long term remote sensing and in-situ studies of key systems • Improved hydrological and ecological models
	Increases in extinctions of marine and terrestrial species • Loss of high percentage of coral reef ecosystems (already underway) • Significant percentage of land mammal, bird, and amphibian species extinct or endangered[7]	Species and population losses accelerating (Portion attributable to climate is uncertain)	**High**	Very high	Moderate	• Better understanding of how species interactions and ecological cascades might magnify extinctions intensity • Better understanding of how interactions between climate-caused extinctions and other extinction drivers (habitat fragmentation, overexploitation, etc.) multiply extinction intensity • Improved monitoring of key species

[1] Change could be either abrupt or non-abrupt.
[2] To clarify, the Committee assesses the near-term outlook that sea level will rise abruptly before the end of this century as Low; this is not in contradiction to the assessment that sea level will continue to rise steadily with estimates of between 0.26 and 0.82 m by the end of this century (IPCC, 2013).
[3] Methane is a powerful but short-lived greenhouse gas.
[4] Limited by ability to predict methane production from thawing organic carbon
[5] No mechanism proposed would lead to abrupt release of substantial amounts of methane from ocean methane hydrates this century.
[6] Limited by uncertainty in hydrate abundance in near-surface sediments, and fate of CH_4 once released
[7] Species distribution models (Thuiller et al., 2006) indicate between 10–40% of mammals now found in African protected areas will be extinct or critically endangered by 2080 as a result of modeled climate change. Analyses by Foden et al.(2013) and Ricke et al. (2013) suggest 41% of bird species, 66% of amphibian species, and between 61% and 100% of corals that are not now considered threatened with extinction will become threatened due to climate change sometime between now and 2100.

References

Abbot, D. S., M. Silber and R. T. Pierrehumbert. 2011. Bifurcations leading to summer Arctic sea ice loss. Journal of Geophysical Research-Atmospheres 116:6017-6027.

ACIA (Arctic Climate Impact Assessment). 2005. Arctic Climate Impact Assessment. Cambridge, UK: Cambridge University Press.

Adams, M. A. 2013. Mega-fires, tipping points and ecosystem services: Managing forests and woodlands in an uncertain future. Forest Ecology and Management 294:250-261.

Aitken, S. N., S. Yeaman, J. A. Holliday, T. L. Wang and S. Curtis-McLane. 2008. Adaptation, migration or extirpation: Climate change outcomes for tree populations. Evolutionary Applications 1(1):95-111.

Alberto, F. J., S. N. Aitken, R. Alía, S. C. González-Martínez, H. Hänninenk, A. Kremer, F. Lefèvre, T. Lenormand, S. Yeaman, R. Whetten and O. Savolainen. 2013. Potential for evolutionary responses to climate change: Evidence from tree populations. Global Change Biology 19:1645-1661.

Alexander, M. A., U. S. Bhatt, J. E. Walsh, M. S. Timlin, J. S. Miller and J. D. Scott. 2004. The atmospheric response to realistic Arctic sea ice anomalies in an AGCM during winter. Journal of Climate 17(5):890-905.

Allen, C. D. and D. D. Breshears. 1998. Drought-induced shift of a forest-woodland ecotone: Rapid landscape response to climate variation. Proceedings of the National Academy of Sciences of the United States of America 95(25):14839-14842.

Allen, M. R., D. J. Frame, C. Huntingford, C. D. Jones, J. A. Lowe, M. Meinshausen and N. Meinshausen. 2009. Warming caused by cumulative carbon emissions towards the trillionth tonne. Nature 458(7242):1163-1166.

Allen, R. J., S. C. Sherwood, J. R. Norris and C. S. Zender. 2012. Recent Northern Hemisphere tropical expansion primarily driven by black carbon and tropospheric ozone. Nature 485(7398):350-393.

Alley, R. B. 2007. Wally was right: Predictive ability of the North Atlantic "conveyor belt" hypothesis for abrupt climate change. Annual Review of Earth and Planetary Sciences 35:241-272.

Alley, R. B., S. Anandakrishnan, T. K. Dupont, B. R. Parizek and D. Pollard. 2007. Effect of sedimentation on ice-sheet grounding-line stability. Science 315(5820):1838-1841.

Alley, R. B., H. J. Horgan, I. Joughin, K. M. Cuffey, T. K. Dupont, B. R. Parizek, S. Anandakrishnan and J. Bassis. 2008. A simple law for ice-shelf calving. Science 322(5906):1344.

Alley, R. B. and I. Joughin. 2012. Modeling Ice-Sheet Flow. Science 336(6081):551-552.

Alley, R. B., J. Marotzke, W. D. Nordhaus, J. T. Overpeck, D. M. Peteet, R. A. Pielke, R. T. Pierrehumbert, P. B. Rhines, T. F. Stocker, L. D. Talley and J. M. Wallace. 2003. Abrupt climate change. Science 299(5615):2005-2010.

Alley, R. B., D. A. Meese, C. A. Shuman, A. J. Gow, K. C. Taylor, P. M. Grootes, J. W. C. White, M. Ram, E. D. Waddington, P. A. Mayewski and G. A. Zielinski. 1993. Abrupt increase in Greenland snow accumulation at the end of the Younger Dryas Event. Nature 362(6420):527-529.

Allison, E. H., A. L. Perry, M. C. Badjeck, W. N. Adger, K. Brown, D. Conway, A. S. Halls, G. M. Pilling, J. D. Reynolds, N. L. Andrew and N. K. Dulvy. 2009. Vulnerability of national economies to the impacts of climate change on fisheries. Fish and Fisheries 10(2):173-196.

Alvarez, L. W., W. Alvarez, F. Asaro and H. V. Michel. 1980. Extraterrestrial cause for the Cretaceous-Tertiary extinction. Science 208(4448):1095-1108.

AMSA (Arctic Marine Shipping Assessment). 2009. Arctic Marine Shipping Assessment 2009 Report. Arctic Council, Arlington, VA.

Anandakrishnan, S., G. A. Catania, R. B. Alley and H. J. Horgan. 2007. Discovery of till deposition at the grounding line of Whillans Ice Stream. Science 315(5820):1835-1838.

Anderegg, W. R. L., J. M. Kane and L. D. L. Anderegg. 2013. Consequences of widespread tree mortality triggered by drought and temperature stress. Nature Climate Change 3(1):30-36.

Andersen, T., J. Carstensen, E. Hernandez-Garcia and C. M. Duarte. 2009. Ecological thresholds and regime shifts: Approaches to identification. Trends in Ecology & Evolution 24(1):49-57.

Anderson-Teixeira, K. J., P. K. Snyder, T. E. Twine, S. V. Cuadra, M. H. Costa and E. H. DeLucia. 2012. Climate-regulation services of natural and agricultural ecoregions of the Americas. Nature Climate Change 2(3):177-181.

Andreae, M. O., C. D. Jones and P. M. Cox. 2005. Strong present-day aerosol cooling implies a hot future. Nature 435(7046):1187-1190.

Andreae, M. O., D. Rosenfeld, P. Artaxo, A. A. Costa, G. P. Frank, K. M. Longo and M. A. F. Silva-Dias. 2004. Smoking rain clouds over the Amazon. Science 303(5662):1337-1342.

Anklin, M., J. M. Barnola, J. Beer, T. Blunier, J. Chappellaz, H. B. Clausen, D. Dahljensen, W. Dansgaard, M. Deangelis, R. J. Delmas, P. Duval, M. Fratta, A. Fuchs, K. Fuhrer, N. Gundestrup, C. Hammer, P. Iversen, S. Johnsen, J. Jouzel, J. Kipfstuhl, M. Legrand, C. Lorius, V. Maggi, H. Miller, J. C. Moore, H. Oeschger, G. Orombelli, D. A. Peel, G. Raisbeck, D. Raynaud, C. Schotthvidberg, J. Schwander, H. Shoji, R. Souchez, B. Stauffer, J. P. Steffensen, M. Stievenard, A. Sveinbjornsdottir, T. Thorsteinsson and E. W. Wolff. 1993. Climate instability during the last interglacial period recorded in the grip ice core. Nature 364(6434):203-207.

AON Design and Implementation Task Force. 2012. Designing, Optimizing, and Implementing an Arctic Observing Network (AON): A Report by the AON Design and Implementation (ADI) Task Force Fairbanks, AK: Study of Environmental Arctic Change (SEARCH).

Arblaster, J. M. and G. A. Meehl. 2006. Contributions of external forcings to southern annular mode trends. Journal of Climate 19(12):2896-2905.

Arblaster, J. M., G. A. Meehl and D. J. Karoly. 2011. Future climate change in the Southern Hemisphere: Competing effects of ozone and greenhouse gases. Geophysical Research Letters 38(2).

Archer, D. 2007. Methane hydrate stability and anthropogenic climate change. Biogeosciences 4(4):521-544.

Archer, D. 2010. The Long Thaw: How Humans Are Changing the Next 100,000 Years of Earth's Climate. Princeton, NJ: Princeton University Press.

Archer, D. and B. Buffett. 2005. Time-dependent response of the global ocean clathrate reservoir to climatic and anthropogenic forcing. Geochemistry Geophysics Geosystems 6(3).

Archer, D., H. Kheshgi and E. MaierReimer. 1997. Multiple timescales for neutralization of fossil fuel CO2. Geophysical Research Letters 24(4):405-408.

Archer, D. E., B. A. Buffett and P. C. McGuire. 2012. A two-dimensional model of the passive coastal margin deep sedimentary carbon and methane cycles. Biogeosciences 9(8):2859-2878.

Armour, K. C., I. Eisenman, E. Blanchard-Wrigglesworth, K. E. McCusker and C. M. Bitz. 2011. The reversibility of sea ice loss in a state-of-the-art climate model. Geophysical Research Letters 38(16).

Arrigo, K. R., G. van Dijken and S. Pabi. 2008. Impact of a shrinking Arctic ice cover on marine primary production. Geophysical Research Letters 35(19).

Arzel, O., T. Fichefet and H. Goosse. 2006. Sea ice evolution over the 20th and 21st centuries as simulated by current AOGCMs. Ocean Modelling 12(3-4):401-415.

Ashwin, P., S. Wieczorek, R. Vitolo and P. Cox. 2012. Tipping points in open systems: bifurcation, noise-induced and rate-dependent examples in the climate system. Philosophical Transactions of the Royal Society A—Mathematical Physical and Engineering Sciences 370(1962):1166-1184.

Asner, G. P., J. K. Clark, J. Mascaro, G. A. G. Garcia, K. D. Chadwick, D. A. N. Encinales, G. Paez-Acosta, E. C. Montenegro, T. Kennedy-Bowdoin, A. Duque, A. Balaji, P. von Hildebrand, L. Maatoug, J. F. P. Bernal, A. P. Y. Quintero, D. E. Knapp, M. C. G. Davila, J. Jacobson and M. F. Ordonez. 2012. High-resolution mapping of forest carbon stocks in the Colombian Amazon. Biogeosciences 9(7):2683-2696.

Asner, G. P., G. V. N. Powell, J. Mascaro, D. E. Knapp, J. K. Clark, J. Jacobson, T. Kennedy-Bowdoin, A. Balaji, G. Paez-Acosta, E. Victoria, L. Secada, M. Valqui and R. F. Hughes. 2010. High-resolution forest carbon stocks and emissions in the Amazon. Proceedings of the National Academy of Sciences of the United States of America 107(38):16738-16742.

Åström, D. O., B. Forsberg, K. L. Ebi and J. Rocklöv. 2013. Attributing mortality from extreme temperatures to climate change in Stockholm, Sweden. Nature Climate Change. DOI:10.1038/nclimate2022.

Augustin, L., C. Barbante, P. R. F. Barnes, J. M. Barnola, M. Bigler, E. Castellano, O. Cattani, J. Chappellaz, D. DahlJensen, B. Delmonte, G. Dreyfus, G. Durand, S. Falourd, H. Fischer, J. Fluckiger, M. E. Hansson, P. Huybrechts, R. Jugie, S. J. Johnsen, J. Jouzel, P. Kaufmann, J. Kipfstuhl, F. Lambert, V. Y. Lipenkov, G. V. C. Littot, A. Longinelli, R. Lorrain, V. Maggi, V. Masson-Delmotte, H. Miller, R. Mulvaney, J. Oerlemans, H. Oerter, G. Orombelli, F. Parrenin, D. A. Peel, J. R. Petit, D. Raynaud, C. Ritz, U. Ruth, J. Schwander, U. Siegenthaler, R. Souchez, B. Stauffer, J. P. Steffensen, B. Stenni, T. F. Stocker, I. E. Tabacco, R. Udisti, R. S. W. van de Wal, M. van den Broeke, J. Weiss, F. Wilhelms, J. G. Winther, E. W. Wolff, M. Zucchelli and E. C. Members. 2004. Eight glacial cycles from an Antarctic ice core. Nature 429(6992):623-628.

Bacmeister, J. T., R. B. Neale, C. Hannay, J. E. Truesdale, J. Caron and M. Wehner. 2013. High-resolution climate simulations using the Community Atmosphere Model (CAM). Journal of Climate, submitted.

Baehr, J., A. Stroup and J. Marotzke. 2009. Testing concepts for continuous monitoring of the meridional overturning circulation in the South Atlantic. Ocean Modelling 29(2):147-153.

Bamber, J. L. and W. P. Aspinall. 2013. An expert judgement assessment of future sea level rise from the ice sheets. Nature Climate Change 3:424-427.

Bamber, J. L., R. E. M. Riva, B. L. A. Vermeersen and A. M. LeBrocq. 2009. Reassessment of the potential sea-level rise from a collapse of the West Antarctic Ice Sheet. Science 324(5929):901-903.

Barnes, E. A. and L. M. Polvani. 2013. Response of the midlatitude jets and of their variability to increased greenhouse gases in the CMIP5 models. Journal of Climate, in press.

Barnosky, A. D. 1986. "Big game" extinction caused by late Pleistocene climatic change: Irish elk (*Megaloceros giganteus*) in Ireland. Quaternary Research 25:128-135.

Barnosky, A. D. 2008. Megafauna biomass tradeoff as a driver of Quaternary and future extinctions. Proceedings of the National Academy of Sciences USA 105(suppl. 1):11543-11548.

Barnosky, A. D. 2009. Heatstroke, Nature in an Age of Global Warming. Washington, DC: Island Press.

Barnosky, A. D., F. Bibi, S. S. B. Hopkins and R. Nichols. 2007. Biostratigraphy and magnetostratigraphy of the mid-Miocene Railroad Canyon Sequence, Montana and Idaho, and age of the mid-Tertiary unconformity west of the continental divide. Journal of Vertebrate Paleontology 27(1):204-224.

Barnosky, A. D., E. A. Hadly, J. Bascompte, E. L. Berlow, J. H. Brown, M. Fortelius, W. M. Getz, J. Harte, A. Hastings, P. A. Marquet, N. D. Martinez, A. Mooers, P. Roopnarine, G. Vermeij, J. W. Williams, R. Gillespie, J. Kitzes, C. Marshall, N. Matzke, D. P. Mindell, E. Revilla and A. B. Smith. 2012. Approaching a state-shift in Earth's biosphere. Nature 486:52-56.

Barnosky, A. D., E. A. Hadly and C. J. Bell. 2003. Mammalian response to global warming on varied temporal scales. Journal of Mammalogy 84(2):354-368.

Barnosky, A. D., P. L. Koch, R. S. Feranec, S. L. Wing and A. B. Shabel. 2004. Assessing the causes of late Pleistocene extinctions on the continents. Science 306:70-75.

Barnosky, A. D. and E. L. Lindsey. 2010. Timing of Quaternary megafaunal extinction in South America in relation to human arrival and climate change. Quaternary International 217:10-29.

Barnosky, A. D., N. Matzke, S. Tomiya, G. O. U. Wogan, B. Swartz, T. Quental, C. Marshall, J. L. McGuire, E. L. Lindsey, K. C. Maguire, B. Mersey and E. A. Ferrer. 2011. Has the Earth's sixth mass extinction already arrived? Nature 471:51-57.

Barton, A., B. Hales, G. G. Waldbusser, C. Langdon and R. A. Feely. 2012. The Pacific oyster, Crassostrea gigas, shows negative correlation to naturally elevated carbon dioxide levels: Implications for near-term ocean acidification effects. Limnology and Oceanography 57(3):698-710.

Bascompte, J. and R. V. Sole. 1996. Habitat fragmentation and extinction thresholds in spatially explicit models. Journal of Animal Ecology 65(4):465-473.

Basher, R. 2006. Global early warning systems for natural hazards: systematic and people-centred. Philosophical Transactions of the Royal Society A—Mathematical Physical and Engineering Sciences 364(1845):2167-2180.

Bathiany, S., M. Claussen and K. Fraedrich. 2012. Implications of climate variability for the detection of multiple equilibria and for rapid transitions in the atmosphere-vegetation system. Climate Dynamics 38(9-10):1775-1790.

Baumann, H., S. C. Talmage and C. J. Gobler. 2011. Reduced early life growth and survival in a fish in direct response to increased carbon dioxide. Nature Climate Change 2(1):38-41.

Becker, A. H., M. Acciaro, R. Asariotis, E. Cabrera, L. Cretegny, P. Crist, M. Esteban, A. Mather, S. Messner, S. Naruse, A. K. Y. Ng, S. Rahmstorf, M. Savonis, D.-W. Song, V. Stenek and A. F. Velegrakis. 2013. A note on climate change adaptation for seaports: A challenge for global ports, a challenge for global society. Climatic Change 120:683-695.

Beever, E. A., C. Ray, J. L. Wilkening, P. F. Brussard and P. W. Mote. 2011. Contemporary climate change alters the pace and drivers of extinction. Global Change Biology 17(6):2054-2070.

Beilman, D. W., G. M. MacDonald, L. C. Smith and P. J. Reimer. 2009. Carbon accumulation in peatlands of West Siberia over the last 2000 years. Global Biogeochemical Cycles 23(1).

Bentz, B. J. 2008. Western U.S. Bark Beetles and Climate Change. U.S. Forest Service Climate Change Resource Center, Washington, DC.

Bentz, B. J., J. Regniere, C. J. Fettig, E. M. Hansen, J. L. Hayes, J. A. Hicke, R. G. Kelsey, J. F. Negron and S. J. Seybold. 2010. Climate Change and Bark Beetles of the Western United States and Canada: Direct and Indirect Effects. Bioscience 60(8):602-613.

Betts, R. A., P. M. Cox, M. Collins, P. P. Harris, C. Huntingford and C. D. Jones. 2004. The role of ecosystem-atmosphere interactions in simulated Amazonian precipitation decrease and forest dieback under global climate warming. Theoretical and Applied Climatology 78(1-3):157-175.

Bevan, S. L., P. R. J. North, W. M. F. Grey, S. O. Los and S. E. Plummer. 2009. Impact of atmospheric aerosol from biomass burning on Amazon dry-season drought. Journal of Geophysical Research-Atmospheres 114(D9).

Biastoch, A., C. W. Boning, J. Getzlaff, J. M. Molines and G. Madec. 2008a. Causes of interannual-decadal variability in the meridional overturning circulation of the midlatitude North Atlantic Ocean. Journal of Climate 21(24):6599-6615.

Biastoch, A., C. W. Boning and J. R. E. Lutjeharms. 2008b. Agulhas leakage dynamics affects decadal variability in Atlantic overturning circulation. Nature 456(7221):489-492.

Bindoff, N. L., J. Willebrand, V. Artale, A. Cazenave, J. Gregory, S. Gulev, K. Hanawa, C. L. Quéré, S. Levitus, Y. Nojiri, C. K. Shum, L. D. Talley and A. Unnikrishnan. 2007. Observations: Oceanic Climate Change and Sea Level. In Climate Change 2007: The Physical Science Basis. Contribution of Working Group I to the Fourth Assessment Report of the Intergovernmental Panel on Climate Change. Solomon, S., D. Qin, M. Manning, Z. Chen, M. Marquis, K. B. Averyt, M. Tignor and H. L. Miller, eds. Cambridge, United Kingdom and New York: Cambridge University Press.

Bindschadler, R. A., S. Nowicki, A. Abe-Ouchi, A. Aschwanden, H. Choi, J. Fastook, G. Granzow, R. Greve, G. Gutowski, U. Herzfeld, C. Jackson, J. Johnson, C. Khroulev, A. Levermann, W. H. Lipscomb, M. A. Martin, M. Morlighem, B. R. Parizek, D. Pollard, S. F. Price, D. D. Ren, F. Saito, T. Sato, H. Seddik, H. Seroussi, K. Takahashi, R. Walker and W. L. Wang. 2013. Ice-sheet model sensitivities to environmental forcing and their use in projecting future sea level (the SeaRISE project). Journal of Glaciology 59(214):195-224.

Bingham, R. J., C. W. Hughes, V. Roussenov and R. G. Williams. 2007. Meridional coherence of the North Atlantic meridional overturning circulation. Geophysical Research Letters 34(23).

Bintanja, R., R. S. W. van de Wal and J. Oerlemans. 2005. Modelled atmospheric temperatures and global sea levels over the past million years. Nature 437(7055):125-128.

Bitz, C. M. and G. H. Roe. 2004. A mechanism for the high rate of sea ice thinning in the Arctic Ocean. Journal of Climate 17(18):3623-3632.

Blanchard-Wrigglesworth, E., K. C. Armour, C. M. Bitz and E. DeWeaver. 2011. Persistence and inherent predictability of Arctic sea ice in a GCM ensemble and observations. Journal of Climate 24(1):231-250.

Blanchard, J. L., S. Jennings, R. Holmes, J. Harle, G. Merino, J. I. Allen, J. Holt, N. K. Dulvy and M. Barange. 2012. Potential consequences of climate change for primary production and fish production in large marine ecosystems. Philosophical Transactions of the Royal Society B—Biological Sciences 367(1605):2979-2989.

Blois, J. L. and E. A. Hadly. 2009. Mammalian response to Cenozoic climatic change. Annual Review of Earth and Planetary Sciences 37(8.1-8.28).

Blois, J. L., P. L. Zarnetske, M. C. Fitzpatrick and S. Finnegan. 2013. Climate change and the past, present, and future of biotic interactions. Science 341(6145):499-504.

Bluhm, B. A. and R. Gradinger. 2008. Regional variability in food availability for arctic marine mammals. Ecological Applications 18(2):S77-S96.

Boden, T. A., G. Marland and R. J. Andres. 2011. Global, Regional, and National Fossil-Fuel CO_2 Emissions. Oak Ridge, TN: Carbon Dioxide Information Analysis Center.

Boe, J. L., A. Hall and X. Qu. 2009. September sea-ice cover in the Arctic Ocean projected to vanish by 2100. Nature Geoscience 2(5):341-343.

Boettiger, C. and A. Hastings. 2012a. Early warning signals and the prosecutor's fallacy. Proceedings of the Royal Society B—Biological Sciences 279(1748):4734-4739.

Boettiger, C. and A. Hastings. 2012b. Quantifying limits to detection of early warning for critical transitions. Journal of the Royal Society Interface 9(75):2527-2539.

Boettiger, C. and A. Hastings. 2013. From patterns to predictions. Nature 493(7431):157-158.

Bouwman, A. F., T. Kram and K. Klein-Goldewijk. 2006. Integrated modelling of global environmental change. An overview of IMAGE 2.4. Bilthoven, The Netherlands: Netherlands Environmental Assessment Agency.

Bower, A. S., M. S. Lozier, S. F. Gary and C. W. Boning. 2009. Interior pathways of the North Atlantic meridional overturning circulation. Nature 459(7244):243-247.

Brewer, P. G., C. Paull, E. T. Peltzer, W. Ussler, G. Rehder and G. Friederich. 2002. Measurements of the fate of gas hydrates during transit through the ocean water column. Geophysical Research Letters 29(22).

Brewer, P. G. and E. T. Peltzer. 2009. Limits to marine life. Science 324(5925):347-348.

Brigham, L. W. 2010. Think again: The Arctic. Foreign Policy 72:70-74.

Brigham, L. W. 2011. Marine protection in the Arctic cannot wait. Nature 478(7368):157.

Brook, B. W. and A. D. Barnosky. 2012. Quaternary extinctions and their link to climate change. In Saving a Million Species. Hannah, L., eds. Washington, DC: Island Press.

Bryan, F. O., G. Danabasoglu, P. R. Gent and K. Lindsay. 2006. Changes in ocean ventilation during the 21st century in the CCSM3. Ocean Modelling 15(3-4):141-156.

Bryden, H. L., H. R. Longworth and S. A. Cunningham. 2005. Slowing of the Atlantic meridional overturning circulation at 25 degrees N. Nature 438(7068):655-657.

Bull, S. R., D. E. Bilello, J. Ekmann, M. J. Sale and D. K. Schmalzer. 2007. Effects of climate change on energy production and distribution in the United States. In Effects of Climate Change on Energy Production and use in the United States. Wilbanks, T. J., V. Bhatt, D. E. Bilello, S. R. Bull, J. Ekmann, W. C. Horak, Y. J. Huang, M. D. Levine, M. J. Sale, D. K. Schmalzer and M. J. Scott, eds. Washington, DC: U.S. Climate Change Science Program.

Bunge, L. and A. J. Clarke. 2009. A verified estimation of the El Nino Index Nino-3.4 since 1877. Journal of Climate 22(14):3979-3992.

Burke, M. B., E. Miguel, S. Satyanath, J. A. Dykema and D. B. Lobell. 2009. Warming increases the risk of civil war in Africa. Proceedings of the National Academy of Sciences of the United States of America 106(49):20670-20674.

Busby, J. W. 2007. Climate Change and National Security: An Agenda for Action. CSR No. 32. New York: Council on Foreign Relations.

Bush, M. B., M. C. Miller, P. E. De Oliveira and P. A. Colinvaux. 2000. Two histories of environmental change and human disturbance in eastern lowland Amazonia. Holocene 10(5):543-553.

Butchart, N., I. Cionni, V. Eyring, T. G. Shepherd, D. W. Waugh, H. Akiyoshi, J. Austin, C. Bruhl, M. P. Chipperfield, E. Cordero, M. Dameris, R. Deckert, S. Dhomse, S. M. Frith, R. R. Garcia, A. Gettelman, M. A. Giorgetta, D. E. Kinnison, F. Li, E. Mancini, C. McLandress, S. Pawson, G. Pitari, D. A. Plummer, E. Rozanov, F. Sassi, J. F. Scinocca, K. Shibata, B. Steil and W. Tian. 2010. Chemistry-climate model simulations of twenty-first century stratospheric climate and circulation changes. Journal of Climate 23(20):5349-5374.

Butchart, N., A. A. Scaife, M. Bourqui, J. de Grandpre, S. H. E. Hare, J. Kettleborough, U. Langematz, E. Manzini, F. Sassi, K. Shibata, D. Shindell and M. Sigmond. 2006. Simulations of anthropogenic change in the strength of the Brewer-Dobson circulation. Climate Dynamics 27(7-8):727-741.

Cahill, A. E., M. E. Aiello-Lammens, M. C. Fisher-Reid, X. Hua, C. J. Karanewsky, H. Y. Ryu, G. C. Sbeglia, F. Spagnolo, J. B. Waldron, O. Warsi, and J. J. Wiens. 2012. How does climate change cause extinction? Proceedings of the Royal Society B—Biological Sciences 280:20121890, http://dx.doi.org/20121810.20121098/rspb.20122012.20121890. http://dx.doi.org/10.1098/rspb.2012.1890.

Cai, W. J., P. H. Whetton and D. J. Karoly. 2003. The response of the Antarctic Oscillation to increasing and stabilized atmospheric CO_2. Journal of Climate 16(10):1525-1538.

Camill, P. 2005. Permafrost thaw accelerates in boreal peatlands during late-20th century climate warming. Climatic Change 68(1-2):135-152.

Cardinale, B. J., J. E. Duffy, A. Gonzalez, D. U. Hooper, C. Perrings, P. Venail, A. Narwani, G. M. Mace, D. Tilman, D. A. Wardle, A. P. Kinzig, G. C. Daily, M. Loreau, J. B. Grace, A. Larigauderie, D. S. Srivastava and S. Naeem. 2012. Biodiversity loss and its impact on humanity. Nature 486:59-67.

Cardinale, B. J., M. A. Palmer and S. L. Collins. 2002. Species diversity enhances ecosystem functioning through interspecific facilitation. Nature 415:426-428.

Carpenter, S. R., J. J. Cole, M. L. Pace, R. Batt, W. A. Brock, T. Cline, J. Coloso, J. R. Hodgson, J. F. Kitchell, D. A. Seekell, L. Smith and B. Weidel. 2011. Early warnings of regime shifts: A whole-ecosystem experiment. Science 332:1079-1082

Centers for Disease Control and Prevention. 1995. Heat-related mortality-Chicago, July 1995, Morbidity and Mortality weekly report,44(11 August):1-4.

Chapin, F. S., G. Peterson, F. Berkes, T. V. Callaghan, P. Angelstam, M. Apps, C. Beier, Y. Bergeron, A. S. Crepin, K. Danell, T. Elmqvist, C. Folke, B. Forbes, N. Fresco, G. Juday, J. Niemela, A. Shvidenko and G. Whiteman. 2004. Resilience and vulnerability of northern regions to social and environmental change. Ambio 33(6):344-349.

Chappellaz, J., T. Blunier, D. Raynaud, J. M. Barnola, J. Schwander and B. Stauffer. 1993. Synchronous changes in atmospheric CH4 and Greenland climate between 40-kyr and 8-kyr BP. Nature 366(6454):443-445.

Checkley, D. M. J., A. G. Dickson, M. Takahashi, J. A. Radich, N. Eisenkolb and R. Asch. 2009. Elevated CO_2 enhances otolith growth in young fish. Science 324:1683.

Chen, I. C., J. K. Hill, R. Ohlemuller, D. B. Roy and C. D. Thomas. 2011. Rapid range shifts of species associated with high levels of climate warming. Science 333(6045):1024-1026.

Chen, Z.-Q. and M. J. Benton. 2012. The timing and pattern of biotic recovery following the end-Permian mass extinction. Nature Geoscience. DOI: 10.1038/NGEO1475.

Chevallier, M. and D. Salas-Melia. 2012. The role of sea ice thickness distribution in the Arctic sea ice potential predictability: A diagnostic approach with a coupled GCM. Journal of Climate 25(8):3025-3038.

Clark, P. U., N. G. Pisias, T. F. Stocker and A. J. Weaver. 2002. The role of the thermohaline circulation in abrupt climate change. Nature 415(6874):863-869.

Clarke, R. A., R. M. Hendry and I. Yashayev. 1998. A western boundary current meter array in the North Atlantic nea 42°N. WOCE Newsletter 33:33-34.

Claussen, M., C. Kubatzki, V. Brovkin, A. Ganopolski, P. Hoelzmann and H. J. Pachur. 1999. Simulation of an abrupt change in Saharan vegetation in the mid-Holocene. Geophysical Research Letters 26(14):2037-2040.

Cochrane, M. A., A. Alencar, M. D. Schulze, C. M. Souza, D. C. Nepstad, P. Lefebvre and E. A. Davidson. 1999. Positive feedbacks in the fire dynamic of closed canopy tropical forests. Science 284(5421):1832-1835.

Collen, B., M. Böhm, R. Kemp, J. E. M. Baillie and (eds.). 2012. Spineless: Status and trends of the world's invertebrates. London: Zoological Society of London.

Collins, M., S. I. An, W. J. Cai, A. Ganachaud, E. Guilyardi, F. F. Jin, M. Jochum, M. Lengaigne, S. Power, A. Timmermann, G. Vecchi and A. Wittenberg. 2010. The impact of global warming on the tropical Pacific ocean and El Nino. Nature Geoscience 3(6):391-397.

Collins, M., R. Knutti, J. Arblaster, J.-L. Dufresne, T. Fichefet, P. Friedlingstein, X. Gao, W. Gutowski, T. Johns, G. Krinner, M. Shongwe, C. Tebaldi, A. Weaver, M. Wehner, M. R. Allen, T. Andrews, U. Beyerle, C. Bitz, S. Bony, B. Booth, O. Brown, V. Brovkin, C. Brutel-Vuilmet, M. Cane, R. Chadwick, E. Cook, K. H. Cook, S. Denvil, M. Eby, J. Fasullo, E. M. Fischer, P. Forster, P. Good, H. Goosse, K. I. Hodges, M. Holland, P. Huybrechts, M. Joshi, V. Kharin, Y. Kushnir, D. Lawrence, R. W. Lee, S. Liddicoat, W. Lucht, D. Matthews, F. Massonnet, M. Meinshausen, C. M. Patricola, G. Philippon-Berthier, Prabhat, S. Rahmstorf, W. J. Riley, J. Rogelj, O. Saenko, R. Seager, J. Sedlacek, L. Shaffrey, D. Shindell, J. Sillmann, A. Slater, R. Webb, G. Zappa and K. Zickfeld. 2012. Long-term climate change: Projections, commitments and irreversibility. In Climate Change 2013: The Physical Science Basis. Contribution of Working Group I to the Fifth Assessment Report of the Intergovernmental Panel on Climate Change. Geneva: World Meteorological Organization.

Confalonieri, U., B. Menne, R. Akhtar, K. L. Ebi, M. Hauengue, R. S. Kovats, B. Revich and A. Woodward. 2007. Human Health. In Climate Change 2007: Impacts, Adaptation and Vulnerability. Contribution of Working Group II to the Fourth Assessment Report of the Intergovernmental Panel on Climate Change. Parry, M. L., O. F. Canziani, J. P. Palutikof, P. J. v. d. Linden and C. E. Hanson, eds. Cambridge, UK: Cambridge University Press.

Cook, B., N. Zeng and J. H. Yoon. 2010a. Climatic and ecological future of the Amazon: Likelihood and causes of change. Earth System Dynamics Discussions 1:63-101.

Cook, E. R., K. J. Anchukaitis, B. M. Buckley, R. D. D'Arrigo, G. C. Jacoby and W. E. Wright. 2010b. Asian monsoon failure and megadrought during the last millennium. Science 328(5977):486-489.

Cook, E. R., R. Seager, R. R. Heim, R. S. Vose, C. Herweijer and C. Woodhouse. 2010c. Megadroughts in North America: Placing IPCC projections of hydroclimatic change in a long-term palaeoclimate context. Journal of Quaternary Science 25(1):48-61.

Cook, P. S. 2011. Impacts of visitor spending on the local economy: Yosemite National Park, 2009 Natural Resource Report NPS: http://www.nps.gov/yose/parkmgmt/upload/YOSE-09-MGM.pdf.

Coope, G. R., A. Morgan and P. J. Osborne. 1971. Fossil Coleoptera as Indicators of climatic fluctuations during last glaciation in Britain. Palaeogeography Palaeoclimatology Palaeoecology 10(2-3):87-101.

Corlett, R. T. and D. A. Westcott. 2013. Will plant movements keep up with climate change? Trends in Ecology & Evolution 28(8):482-488.

Costa, M. H. and G. F. Pires. 2010. Effects of Amazon and Central Brazil deforestation scenarios on the duration of the dry season in the arc of deforestation. International Journal of Climatology 30(13):1970-1979.

Cox, P. M., R. A. Betts, C. D. Jones, S. A. Spall and I. J. Totterdell. 2000. Acceleration of global warming due to carbon-cycle feedbacks in a coupled climate model. Nature 408(6809):184-187.

Cronin, T. M. 2012. Rapid sea-level rise. Quaternary Science Reviews 56:11-30.

Cunningham, S. A., T. Kanzow, D. Rayner, M. O. Baringer, W. E. Johns, J. Marotzke, H. R. Longworth, E. M. Grant, J. J. M. Hirschi, L. M. Beal, C. S. Meinen and H. L. Bryden. 2007. Temporal variability of the Atlantic meridional overturning circulation at 26.5 degrees N. Science 317(5840):935-938.

Cunningham, S. A. and R. Marsh. 2010. Observing and modeling changes in the Atlantic MOC. Wiley Interdisciplinary Reviews—Climate Change 1(2):180-191.

D'Odorico, P., A. Bhattachan, K. F. Davis, S. Ravi and C. W. Runyan. 2013. Global desertification: Drivers and feedbacks. Advances in Water Resources 51:326-344.

da Rocha, H. R., A. O. Manzi, O. M. Cabral, S. D. Miller, M. L. Goulden, S. R. Saleska, N. R. Coupe, S. C. Wofsy, L. S. Borma, P. Artaxo, G. Vourlitis, J. S. Nogueira, F. L. Cardoso, A. D. Nobre, B. Kruijt, H. C. Freitas, C. von Randow, R. G. Aguiar and J. F. Maia. 2009. Patterns of water and heat flux across a biome gradient from tropical forest to savanna in Brazil. Journal of Geophysical Research—Biogeosciences 114(G1).

Dai, A. G. 2011. Drought under global warming: A review. Wiley Interdisciplinary Reviews-Climate Change 2(1):45-65.

Daily, G. C. and K. Ellison. 2002. The New Economy of Nature: The Quest to Make Conservation Profitable. Washington, DC: Island Press.

Daily, G. C., T. Soderqvist, S. Aniyar, K. Arrow, P. Dasgupta, P. R. Ehrlich, C. Folke, A. Jansson, B. O. Jansson, N. Kautsky, S. Levin, J. Lubchenco, K. G. Maler, D. Simpson, D. Starrett, D. Tilman and B. Walker. 2000. Ecology—The value of nature and the nature of value. Science 289(5478):395-396.

Dakos, V., M. Scheffer, E. H. van Nes, V. Brovkin, V. Petoukhov and H. Held. 2008. Slowing down as an early warning signal for abrupt climate change. Proceedings of the National Academy of Sciences of the United States of America 105(38):14308-14312.

Danabasoglu, G., S. G. Yeager, D. Bailey, E. Behrens, M. Bentsen, D. Bi, A. Biastoch, C. Böning, A. Bozec, V. M. Canuto, C. Cassou, E. Chassignet, A. C. Coward, S. Danilov, N. Diansky, H. Drange, R. Farneti, E. Fernandez, P. G. Fogli, G. Forget, Y. Fujii, S. M. Griffies, A. Gusev, P. Heimbach, A. Howard, T. Jung, M. Kelley, W. G. Large, A. Leboissetier, J. Lu, G. Madec, S. J. Marsland, S. Masina, A. Navarra, A. J. G. Nurser, A. Pirani, D. S. y. M´elia, B. L. Samuels, M. Scheinert, D. Sidorenko, A.-M. Treguier, H. Tsujino, P. Uotila, S. Valcke, A. Voldoire and Q. Wang. 2013. North Atlantic Simulations in Coordinated Ocean-ice Reference Experiments phase II (CORE-II). Part I: Mean States. Preprint submitted to Ocean Modelling.

Danise, S., R. J. Twitchett, C. T. S. Little and M. E. Clemence. 2013. The impact of global warming and anoxia on marine benthic community dynamics: An example from the Toarcian (Early Jurassic). PLoS ONE 8(2).

Dansgaard, W., J. W. C. White and S. J. Johnsen. 1989. The abrupt termination of the Younger Dryas Climate Event. Nature 339(6225):532-534.

Davie, M. K. and B. A. Buffett. 2001. A numerical model for the formation of gas hydrate below the seafloor. Journal of Geophysical Research-Solid Earth 106(B1):497-514.

Davis, S. J., L. Cao, K. Caldeira and M. I. Hoffert. 2013. Rethinking wedges. Environmental Research Letters 8(1).

Davis, S. M. and K. H. Rosenlof. 2012. A multidiagnostic intercomparison of tropical-width time series using reanalyses and satellite observations. Journal of Climate 25(4):1061-1078.

Dawson, T. P., S. T. Jackson, J. I. House, I. C. Prentice and G. M. Mace. 2011. Beyond predictions: Biodiversity conservation in a changing climate. Science 332:53-58.

De Costa, B. F. 1879. The Lenox Globe. Magazine of American History 3(9):12.

de Vries, P. and S. L. Weber. 2005. The Atlantic freshwater budget as a diagnostic for the existence of a stable shut down of the meridional overturning circulation. Geophysical Research Letters 32(9).

Delaygue, G., E. Bard, C. Rollion, J. Jouzel, M. Stievenard, J. C. Duplessy and G. Ganssen. 2001. Oxygen isotope/salinity relationship in the northern Indian Ocean. Journal of Geophysical Research-Oceans 106(C3):4565-4574.

Delworth, T. L., P. U. Clark, M. Holland, W. E. Johns, T. Kuhlbrodt, J. Lynch-Stieglitz, C. Morrill, R. Seager, A. J. Weaver and R. Zhang. 2008. The potential for abrupt change in the Atlantic Meridional Overturning Circulation. In Abrupt Climate Change: A report by the U.S. Climate Change Science Program and the Subcommittee on Global Change Research, eds. Reston, VA: U.S. Geological Survey.

Denman, K. L., G. Brasseur, A. Chidthaisong, P. Ciais, P. M. Cox, R. E. Dickinson, D. Hauglustaine, C. Heinze, E. Holland, D. Jacob, U. Lohmann, S. Ramachandran, P. L. d. S. Dias, S. C. Wofsy and X. Zhang. 2007. Couplings between changes in the climate system and biogeochemistry. In Climate Change 2007: The Physical Science Basis. Contribution of Working Group I to the Fourth Assessment Report of the Intergovernmental Panel on Climate Change. Solomon, S., D. Qin, M. Manning, Z. Chen, M. Marquis, K. B. Averyt, M. Tignor and H. L. Miller, eds. Cambridge, UK and New York: Cambridge University Press.

Derocher, A. E., N. J. Lunn and I. Stirling. 2004. Polar bears in a warming climate. Integrative and Comparative Biology 44(2):163-176.

Deser, C., R. Knutti, S. Solomon and A. S. Phillips. 2012a. Communication of the role of natural variability in future North American climate. Nature Climate Change 2(11):775-779.

Deser, C., A. Phillips, V. Bourdette and H. Y. Teng. 2012b. Uncertainty in climate change projections: the role of internal variability. Climate Dynamics 38(3-4):527-546.

Deser, C., R. Tomas, M. Alexander and D. Lawrence. 2010. The seasonal atmospheric response to projected arctic sea ice loss in the late twenty-first century. Journal of Climate 23(2):333-351.

Deutsch, C., H. Brix, T. Ito, H. Frenzel and L. Thompson. 2011. Climate-forced variability of ocean hypoxia. Science 333(6040):336-339.

Diaz, R. J. and R. Rosenberg. 2008. Spreading dead zones and consequences for marine ecosystems. Science 321(5891):926-929.

Diffenbaugh, N. S. and C. B. Field. 2013. Changes in ecologically critical terrestrial climate conditions. Science 341(6145):486-492.

Dijkstra, H. A. 2007. Characterization of the multiple equilibria regime in a global ocean model. Tellus Series A—Dynamic Meteorology and Oceanography 59(5):695-705.

DiNezio, P. N., G. A. Vecchi and A. C. Clement. 2013. Detectability of changes in the Walker circulation in response to global warming. Journal of Climate 26(12):4038-4048.

Dirzo, R. and P. H. Raven. 2003. Global state of biodiversity and loss. Annual Review of Environment and Resources 28:137-167.

Ditlevsen, P. D. and S. J. Johnsen. 2010. Tipping points: Early warning and wishful thinking. Geophysical Research Letters 37(19).

Dow, K. and T. E. Downing. 2007. The Atlas of Climate Change. Berkeley: University of California Press.

Dowdeswell, J. A., D. Ottesen, J. Evans, C. O. Cofaigh and J. B. Anderson. 2008. Submarine glacial landforms and rates of ice-stream collapse. Geology 36(10):819-822.

Drake, J. M. and B. D. Griffen. 2010. Early warning signals of extinction in deteriorating environments. Nature 467:456-459.

Driesschaert, E., T. Fichefet, H. Goosse, P. Huybrechts, I. Janssens, A. Mouchet, G. Munhoven, V. Brovkin and S. L. Weber. 2007. Modeling the influence of Greenland ice sheet melting on the Atlantic meridional overturning circulation during the next millennia. Geophysical Research Letters 34(10).

Drijfhout, S. S., S. L. Weber and E. van der Swaluw. 2011. The stability of the MOC as diagnosed from model projections for pre-industrial, present and future climates. Climate Dynamics 37(7-8):1575-1586.

Dunne, J. P., R. J. Stouffer and J. G. John. 2013. Reductions in labour capacity from heat stress under climate warming. Nature Climate Change 3:563-566.

Durner, G. M., D. C. Douglas, R. M. Nielson, S. C. Amstrup, T. L. McDonald, I. Stirling, M. Mauritzen, E. W. Born, O. Wiig, E. DeWeaver, M. C. Serreze, S. E. Belikov, M. M. Holland, J. Maslanik, J. Aars, D. A. Bailey and A. E. Derocher. 2009. Predicting 21st-century polar bear habitat distribution from global climate models. Ecological Monographs 79(1):25-58.

Early, R. and D. Sax. 2011. Analysis of climate paths reveals potential limitations on species range shifts. Ecology Letters 14:1125-1133.

Easterling, W. E., P. Aggarwal, P. Batima, K. M. Brander, L. Erda, S. M. Howden, A. Kirilenko, J. Morton, J.-F. Soussana, J. Schmidhuber and F. Tubiello. 2012. Food, fibre and forest products. In Climate Change 2007: Impacts, Adaptation and Vulnerability. Contribution of Working Group II to the Fourth Assessment Report of the Intergovernmental Panel on Climate Change. Parry, M., O. Canziani, J. Palutikof, P. V. D. Linden and C. Hanson, eds. Cambridge, UK: Cambridge University Press.

EEA (European Environment Agency). 2001. Late lessons from early warnings: the precautionary principle 1896-2000. Environmental Issue Report. Copenhagen, Denmark: European Environment Agency.

EEA. 2013. Late lessons from early warnings: science, precaution, innovation. Copenhagen, Denmark: European Environment Agency.

Ehrlich, P. R., P. M. Kareiva and G. C. Daily. 2012. Securing natural capital and expanding equity to rescale civilization. Nature 486(7401):68-73.

Eisenman, I. 2012. Factors controlling the bifurcation structure of sea ice retreat. Journal of Geophysical Research—Atmospheres 117(D1).

Eisenman, I. and J. S. Wettlaufer. 2009. Nonlinear threshold behavior during the loss of Arctic sea ice. Proceedings of the National Academy of Sciences of the United States of America 106(1):28-32.

Eltahir, E. A. B. and R. L. Bras. 1994. Precipitation recycling in the Amazon Basin. Quarterly Journal of the Royal Meteorological Society 120(518):861-880.

Famiglietti, J. S., M. Lo, S. L. Ho, J. Bethune, K. J. Anderson, T. H. Syed, S. C. Swenson, C. R. de Linage and M. Rodell. 2011. Satellites measure recent rates of groundwater depletion in California's Central Valley. Geophysical Research Letters 38(3).

Farman, J. C., B. G. Gardiner and J. D. Shanklin. 1985. Large losses of total ozone in antarctica reveal seasonal Clox/Nox interaction. Nature 315(6016):207-210.

Fearnside, P. M. 1983. Land-use trends in the Brazilian Amazon region as factors in accelerating deforestation. Environmental Conservation 10(2):141-148.

Feldstein, S. B. 2000. Teleconnections and ENSO: The timescale, power spectra, and climate noise properties. Journal of Climate 13(4430):4430-4440.

Ferrari, M. C. O., R. P. Manassa, D. L. Dixson, P. L. Munday, M. I. McCormick, M. G. Meekan, A. Sih and D. P. Chivers. 2012. Effects of ocean acidification on learning in coral reef fishes. PLoS ONE 7(2): e31478..

Fetterer, F., K. Knowles, W. Meier and M. Savoie. 2012. Sea ice index. Accessed at http://nsidc.org/data/seaice_index/.

Field, C. B., V. Barros, T. F. Stocker, D. Qin, D. J. Dokken, K. L. Ebi, M. D. Mastrandrea, K. J. Mach, G.-K. Plattner, S. K. Allen, M. Tignor and P. M. Midgley, Eds. 2012. Managing the Risks of Extreme Events and Disasters to Advance Climate Change Adaptation. A Special Report of Working Groups I and II of the Intergovernmental Panel on Climate Change (IPCC). Cambridge, UK and New York: Cambridge University Press.

Fingar, T. 2008. National Security Implications of Global Climate Change to 2030 Testimony to the U.S. House of Representatives Permanent Select Committee on Intelligence and Select Committee on Energy Independence and Global Warming. Hearing, June 25, 2008.

Fischer, J. and F. A. Schott. 2002. Labrador Sea water tracked by profiling floats—From the boundary current into the open North Atlantic. Journal of Physical Oceanography 32(2):573-584.

Flato, G. M. and R. D. Brown. 1996. Variability and climate sensitivity of landfast Arctic sea ice. Journal of Geophysical Research—Oceans 101(C11):25767-25777.

Foden, W. B., S. H. M. Butchart, S. N. Stuart, J.-C. Vié, H. R. Akçakaya, A. Angulo, L. M. DeVantier, A. Gutsche, E. Turak, L. Cao, S. D. Donner, V. Katariya, R. Bernard, R. A. Holland, A. F. Hughes, S. E. O'Hanlon, S. T. Garnett, Ç. H. Şekercioğlu and G. M. Mace. 2013. Identifying the world's most climate change vulnerable species: A systematic trait-based assessment of all birds, amphibians and corals. PLoS ONE 8(6):e65427.

Foley, J. A., N. Ramankutty, K. A. Brauman, E. S. Cassidy, J. S. Gerber, M. Johnston, N. D. Mueller, C. O'Connell, D. K. Ray, P. C. West, C. Balzer, E. M. Bennett, S. R. Carpenter, J. Hill, C. Monfreda, S. Polasky, J. Rockström, J. Sheehan, S. Siebert, D. Tilman and D. P. M. Zaks. 2011. Solutions for a cultivated planet. Nature 478:337-342.

Francis, J. A. and S. J. Vavrus. 2012. Evidence linking Arctic amplification to extreme weather in mid-latitudes. Geophysical Research Letters 39(6).

Fretwell, P., H. D. Pritchard, D. G. Vaughan, J. L. Bamber, N. E. Barrand, R. Bell, C. Bianchi, R. G. Bingham, D. D. Blankenship, G. Casassa, G. Catania, D. Callens, H. Conway, A. J. Cook, H. F. J. Corr, D. Damaske, V. Damm, F. Ferraccioli, R. Forsberg, S. Fujita, Y. Gim, P. Gogineni, J. A. Griggs, R. C. A. Hindmarsh, P. Holmlund, J. W. Holt, R. W. Jacobel, A. Jenkins, W. Jokat, T. Jordan, E. C. King, J. Kohler, W. Krabill, M. Riger-Kusk, K. A. Langley, G. Leitchenkov, C. Leuschen, B. P. Luyendyk, K. Matsuoka, J. Mouginot, F. O. Nitsche, Y. Nogi, O. A. Nost, S. V. Popov, E. Rignot, D. M. Rippin, A. Rivera, J. Roberts, N. Ross, M. J. Siegert, A. M. Smith, D. Steinhage, M. Studinger, B. Sun, B. K. Tinto, B. C. Welch, D. Wilson, D. A. Young, C. Xiangbin and A. Zirizzotti. 2013. Bedmap2: Improved ice bed, surface and thickness datasets for Antarctica. Cryosphere 7(1):375-393.

Friedlingstein, P., P. Cox, R. Betts, L. Bopp, W. Von Bloh, V. Brovkin, P. Cadule, S. Doney, M. Eby, I. Fung, G. Bala, J. John, C. Jones, F. Joos, T. Kato, M. Kawamiya, W. Knorr, K. Lindsay, H. D. Matthews, T. Raddatz, P. Rayner, C. Reick, E. Roeckner, K. G. Schnitzler, R. Schnur, K. Strassmann, A. J. Weaver, C. Yoshikawa and N. Zeng. 2006. Climate-carbon cycle feedback analysis: Results from the (CMIP)-M-4 model intercomparison. Journal of Climate 19(14):3337-3353.

Friedlingstein, P., R. A. Houghton, G. Marland, J. Hackler, T. A. Boden, T. J. Conway, J. G. Canadell, M. R. Raupach, P. Ciais and C. Le Quere. 2010. Update on CO_2 emissions. Nature Geoscience 3(12):811-812.

Fu, Q., C. M. Johanson, J. M. Wallace and T. Reichler. 2006. Enhanced mid-latitude tropospheric warming in satellite measurements. Science 312(5777):1179.

Fu, R. and W. Li. 2004. The influence of the land surface on the transition from dry to wet season in Amazonia. Theoretical and Applied Climatology 78(1-3):97-110.

Fyfe, J. C., G. J. Boer and G. M. Flato. 1999. The Arctic and Antarctic oscillations and their projected changes under global warming. Geophysical Research Letters 26(11):1601-1604.

Garcia, R. R. and W. J. Randel. 2008. Acceleration of the Brewer-Dobson circulation due to increases in greenhouse gases. Journal of the Atmospheric Sciences 65(8):2731-2739.

Garny, H., M. Dameris, W. Randel, G. E. Bodeker and R. Deckert. 2011. Dynamically forced increase of tropical upwelling in the lower stratosphere. Journal of the Atmospheric Sciences 68(6):1214-1233.

Gary, S. F., M. S. Lozier, C. W. Boning and A. Biastoch. 2011. Deciphering the pathways for the deep limb of the Meridional Overturning Circulation. Deep-Sea Research Part II—Topical Studies in Oceanography 58(17-18):1781-1797.

Gash, J. H. C. and C. A. Nobre. 1997. Climatic effects of Amazonian deforestation: Some results from ABRACOS. Bulletin of the American Meteorological Society 78(5):823-830.

Gasson, E., M. Siddall, D. J. Lunt, O. J. L. Rackham, C. H. Lear and D. Pollard. 2012. Exploring uncertainties in the relationship between temperature, ice volume, and sea level over the past 50 million years. Reviews of Geophysics 50(1).

Gazeau, F., J. P. Gattuso, M. Greaves, H. Elderfield, J. Peene, C. H. R. Heip and J. J. Middelburg. 2011. Effect of carbonate chemistry alteration on the early embryonic development of the Pacific Oyster (*Crassostrea gigas*). PLoS ONE 6(8):e23010.

Gedan, K. B., M. L. Kirwan, E. Wolanski, E. B. Barbier and B. R. Silliman. 2011. The present and future role of coastal wetland vegetation in protecting shorelines: answering recent challenges to the paradigm. Climatic Change 106(1):7-29.

Giles, K. A., S. W. Laxon and A. L. Ridout. 2008. Circumpolar thinning of Arctic sea ice following the 2007 record ice extent minimum. Geophysical Research Letters 35(22).

Gillett, N.P. and D.W.J. Thompson. 2003. Simulation of recent Southern Hemisphere climate change. Science 302(5643):273-275.

Gloor, M., L. Gatti, R. Brienen, T. R. Feldpausch, O. L. Phillips, J. Miller, J. P. Ometto, H. Rocha, T. Baker, B. de Jong, R. A. Houghton, Y. Malhi, L. E. O. C. Aragao, J. L. Guyot, K. Zhao, R. Jackson, P. Peylin, S. Sitch, B. Poulter, M. Lomas, S. Zaehle, C. Huntingford, P. Levy and J. Lloyd. 2012. The carbon balance of South America: A review of the status, decadal trends and main determinants. Biogeosciences 9(12):5407-5430.

Godfray, H. C. J., J. R. Beddington, I. R. Crute, L. Haddad, D. Lawrence, J. F. Muir, J. Pretty, S. Robinson, S. M. Thomas and C. Toulmin. 2010. Food security: The challenge of feeding 9 billion people. Science 327(5967):812-818.

Goetz, S. J., M. C. Mack, K. R. Gurney, J. T. Randerson and R. A. Houghton. 2007. Ecosystem responses to recent climate change and fire disturbance at northern high latitudes: Observations and model results contrasting northern Eurasia and North America. Environmental Research Letters 2(4).

Golding, N. and R. Betts. 2008. Fire risk in Amazonia due to climate change in the HadCM3 climate model: Potential interactions with deforestation. Global Biogeochemical Cycles 22(4).

Good, P., C. Jones, J. Lowe, R. Betts and N. Gedney. 2013. Comparing tropical forest projections from two generations of Hadley Centre Earth System Models, HadGEM2-ES and HadCM3LC. Journal of Climate 26(2):495-511.

Goodess, C. M., C. Hanson, M. H. M and T. J. Osborn. 2003. Representing climate and extreme weather events in integrated assessment models: a review of existing methods and options for development. Integrated Assessment 4:145-171.

Graham, A. 1999. Late Cretaceous and Cenozoic History of North American Vegetation. New York: Oxford University Press.

Grayson, D. K. 2005. A brief history of Great Basin pikas. Journal of Biogeography 32(12):2103-2111.

Gregory, J. M., K. W. Dixon, R. J. Stouffer, A. J. Weaver, E. Driesschaert, M. Eby, T. Fichefet, H. Hasumi, A. Hu, J. H. Jungclaus, I. V. Kamenkovich, A. Levermann, M. Montoya, S. Murakami, S. Nawrath, A. Oka, A. P. Sokolov and R. B. Thorpe. 2005. A model intercomparison of changes in the Atlantic thermohaline circulation in response to increasing atmospheric CO_2 concentration. Geophysical Research Letters 32(12).

Grootes, P. M., M. Stuiver, J. W. C. White, S. Johnsen and J. Jouzel. 1993. Comparison of oxygen-isotope records from the Gisp2 and Grip Greenland ice cores. Nature 366(6455):552-554.

Grundmann, R. 2002. Transnational Environmental Policy: Reconstructing Ozone. London and New York: Routledge.

Hadly, E. A. and A. D. Barnosky. 2009. Vertebrate fossils and the future of conservation biology. In Conservation Paleobiology: Using the Past to Manage for the Future, The Paleontological Society Papers vol. 15. Dietl, G. P. and K. W. Flessa, eds. New Haven, CT: Yale University Printing and Publishing Services.

Hadly, E. A., U. Ramakrishnan, Y. L. Chan, M. van Tuinen, K. O'Keefe, P. A. Spaeth and C. J. Conroy. 2004. Genetic response to climatic change: Insights from ancient DNA and phylochronology. PLoS Biology 2(10):1600-1609.

Halpern, B. S., S. Walbridge, K. A. Selkoe, C. V. Kappel, F. Micheli, C. D'Agrosa, J. F. Bruno, K. S. Casey, C. Ebert, H. E. Fox, R. Fujita, D. Heinemann, H. S. Lenihan, E. M. P. Madin, M. T. Perry, E. R. Selig, M. Spalding, R. Steneck and R. Watson. 2008. A global map of human impact on marine ecosystems. Science 319:948-952.

Hansen, J., M. Sato and R. Ruedy. 2012. Perception of climate change. Proceedings of the National Academy of Sciences of the United States of America 109(37):E2415-E2423.

Hansen, J., M. Sato and R. Ruedy. 2013a. Global Temperature Update Through 2012. Available at: http://www.columbia.edu/~jeh1/mailings/2013/20130115_Temperature2012.pdf; accessed August 14, 2013.

Hansen, J., M. Sato and R. Ruedy. 2013b. Reply to Rhines and Huybers: Changes in the frequency of extreme summer heat. Proceedings of the National Academy of Sciences of the United States of America 110(7):E547-E548.

Harnik, P. G., H. K. Lotze, S. C. Anderson, Z. V. Finkel, S. Finnegan, D. R. Lindberg, L. H. Liow, R. Lockwood, C. R. McClain, J. L. McGuire, A. O'Dea, J. M. Pandolfi, C. Simpson and D. P. Tittensor. 2012. Extinctions in ancient and modern seas. Trends in Ecology & Evolution 27(11).

Hartmann, D. L. and F. Lo. 1998. Wave-driven zonal flow vacillation in the Southern Hemisphere. Journal of the Atmospheric Sciences 55(8):1303-1315.

Hastings, A. and D. Wysham. 2010. Regime shifts in ecological systems can occur with no warning. Ecology Letters 13:464-472.

Hawkins, E., R. S. Smith, L. C. Allison, J. M. Gregory, T. J. Woollings, H. Pohlmann and B. de Cuevas. 2011. Bistability of the Atlantic overturning circulation in a global climate model and links to ocean freshwater transport. Geophysical Research Letters 38(L10605).

Hays, J. D., J. Imbrie and N. J. Shackleton. 1976. Variations in Earth's orbit: Pacemaker of the ice ages. Science 194:1121-1132.

He, F., J. D. Shakun, P. U. Clark, A. E. Carlson, Z. Y. Liu, B. L. Otto-Bliesner and J. E. Kutzbach. 2013. Northern Hemisphere forcing of Southern Hemisphere climate during the last deglaciation. Nature 494(7435):81-85.

Headly, M. A. and J. P. Severinghaus. 2007. A method to measure Kr/N-2 ratios in air bubbles trapped in ice cores and its application in reconstructing past mean ocean temperature. Journal of Geophysical Research—Atmospheres 112(D19).

Held, H. and T. Kleinen. 2004. Detection of climate system bifurcations by degenerate fingerprinting. Geophysical Research Letters 31(23).

Helm, K. P., N. L. Bindoff and J. A. Church. 2011. Observed decreases in oxygen content of the global ocean. Geophysical Research Letters 38(23).

Hidalgo, M., T. Rouyer, V. Bartolino, S. Cervino, L. Ciannelli, E. Massuti, A. Jadaud, F. Saborido-Rey, J. M. Durant, M. Santurtun, C. Pineiro and N. C. Stenseth. 2012. Context-dependent interplays between truncated demographies and climate variation shape the population growth rate of a harvested species. Ecography 35(7):637-649.

Hill, J. C., N. W. Driscoll, J. K. Weissel and J. A. Goff. 2004. Large-scale elongated gas blowouts along the US Atlantic margin. Journal of Geophysical Research—Solid Earth 109(B9).

Hinzman, L. D., N. D. Bettez, W. R. Bolton, F. S. Chapin, M. B. Dyurgerov, C. L. Fastie, B. Griffith, R. D. Hollister, A. Hope, H. P. Huntington, A. M. Jensen, G. J. Jia, T. Jorgenson, D. L. Kane, D. R. Klein, G. Kofinas, A. H. Lynch, A. H. Lloyd, A. D. McGuire, F. E. Nelson, W. C. Oechel, T. E. Osterkamp, C. H. Racine, V. E. Romanovsky, R. S. Stone, D. A. Stow, M. Sturm, C. E. Tweedie, G. L. Vourlitis, M. D. Walker, D. A. Walker, P. J. Webber, J. M. Welker, K. Winker and K. Yoshikawa. 2005. Evidence and implications of recent climate change in northern Alaska and other arctic regions. Climatic Change 72(3):251-298.

Hirota, M., M. Holmgren, E. H. Van Nes and M. Scheffer. 2011. Global resilience of tropical forest and savanna to critical transitions. Science 334(6053):232-235.

Hodell, D. A., J. H. Curtis and M. Brenner. 1995. Possible role of climate in the collapse of classic Maya civilization. Nature 375(6530):391-394.

Hoegh-Guldberg, O. 1999. Climate change, coral bleaching, and the future of the world's coral reefs. Marine and Freshwater Research 50:839-866.

Hoegh-Guldberg, O., L. Hughes, S. McIntyre, D. B. Lindenmayer, C. Parmesan, H. P. Possingham and C. D. Thomas. 2008. Assisted colonization and rapid climate change. Science 321(5887):345-346.

Hoekstra, J. M., J. L. Molnar, M. Jennings, C. Revenga, M. D. Spaulding, T. M. Boucher, J. C. Robertson, T. J. Heibel and K. Ellison. 2010. The Atlas of Global Conservation. Berkeley: University of California Press.

Hoffmann, A. A. and C. M. Sgro. 2011. Climate change and evolutionary adaptation. Nature 470(7335):479-485.

Hofmann, A. F., E. T. Peltzer, P. M. Walz and P. G. Brewer. 2011. Hypoxia by degrees: Establishing definitions for a changing ocean. Deep-Sea Research Part I—Oceanographic Research Papers 58(12):1212-1226.

Hofmann, M. and S. Rahmstorf. 2009. On the stability of the Atlantic meridional overturning circulation. Proceedings of the National Academy of Sciences of the United States of America 106(49):20584-20589.

Holdo, R. M., R. D. Holt and J. M. Fryxell. 2009. Grazers, browsers, and fire influence the extent and spatial pattern of tree cover in the Serengeti. Ecological Applications 19(1):95-109.

Holdridge, L. R. 1947. Determination of world plant formations from simple climatic data. Science 105(2727):367-368.

Holdridge, L. R. 1964. Life Zone Ecology. San Jose, Costa Rica: Tropical Science Center.

Holland, M. M., D. A. Bailey and S. Vavrus. 2011. Inherent sea ice predictability in the rapidly changing Arctic environment of the Community Climate System Model, version 3. Climate Dynamics 36(7-8):1239-1253.

Holland, M. M., C. M. Bitz and B. Tremblay. 2006. Future abrupt reductions in the summer Arctic sea ice. Geophysical Research Letters 33(23).

Holland, M. M., C. M. Bitz, B. Tremblay and D. A. Bailey. 2008. The role of natural versus forced change in future rapid summer Arctic ice loss. In Arctic Sea Ice Decline: Observations, Projections, Mechanisms, and Implications. Geophysical Monograph Series, Vol. 180. DeWeaver, E. T., C. M. Bitz and L.-B. Tremblay, eds. Washington, DC: American Geophysical Union.

Honey, M. 2008. Ecotourism and Sustainable Development, Second Edition: Who Owns Paradise? Washington, DC: Island Press.

Hönisch, B., A. Ridgwell, D. N. Schmidt, E. Thomas, S. J. Gibbs, A. Sluijs, R. Zeebe, L. Kump, R. C. Martindale, S. E. Greene, W. Kiessling, J. Ries, J. C. Zachos, D. L. Royer, S. Barker, J. Thomas M. Marchitto, R. Moyer, C. Pelejero, P. Ziveri, G. L. Foster and B. Williams. 2012. The geological record of ocean acidification. Science 335:1058-1063.

Horgan, H. J. and S. Anandakrishnan. 2006. Static grounding lines and dynamic ice streams: Evidence from the Siple Coast, West Antarctica. Geophysical Research Letters 33(18).

Houghton, R. A. 2008. TRENDS: A Compendium of Data on Global Change Carbon Dioxide. Oak Ridge, TN: Information Analysis Center, Oak Ridge National Laboratory, US Department of Energy.

Hsiang, S. M., M. Burke and E. Miguel. 2013. Quantifying the influence of climate on human conflict. Science 341(6151).

Hsiang, S. M., K. C. Meng and M. A. Cane. 2011. Civil conflicts are associated with the global climate. Nature 476(7361):438-441.

Hu, A. X., G. A. Meeh, W. Q. Han and J. J. Yin. 2009. Transient response of the MOC and climate to potential melting of the Greenland Ice Sheet in the 21st century. Geophysical Research Letters 36(10).

Huey, R. B., M. R. Kearney, A. Krockenberger, J. A. M. Holtum, M. Jess and S. E. Williams. 2012. Predicting organismal vulnerability to climate warming: roles of behaviour, physiology and adaptation. Philosophical Transactions of the Royal Society B—Biological Sciences 367(1596):1665-1679.

Hughes, S., A. Yau, L. Max, N. Petrovic, F. Davenport, M. Marshall, T. R. McClanahan, E. H. Allison and J. E. Cinner. 2012. A framework to assess national level vulnerability from the perspective of food security: The case of coral reef fisheries. Environmental Science & Policy 23:95-108.

Huisman, S. E., M. den Toom, H. A. Dijkstra and S. Drijfhout. 2010. An indicator of the multiple equilibria regime of the Atlantic meridional overturning circulation. Journal of Physical Oceanography 40(3):551-567.

Huntington, H. P., E. Goodstein and E. Euskirchen. 2012. Towards a tipping point in responding to change: Rising costs, fewer options for Arctic and global societies. Ambio 41(1):66-74.

Hurrell, J. W., M. M. Holland, S. Ghan, J.-F. Lamarque, D. Lawrence, W. H. Lipscomb, N. Mahowald, D. Marsh, P. Rasch, D. Bader, W. D. Collins, P. R. Gent, J. J. Hack, J. Kiehl, P. Kushner, W. G. Large, S. Marshall, S. Vavrus and M. Vertenstein. 2013. The Community Earth System Model: A Framework for Collaborative Research. Bulletin of the American Meteorological Society, in press.

Hutyra, L. R., J. W. Munger, C. A. Nobre, S. R. Saleska, S. A. Vieira and S. C. Wofsy. 2005. Climatic variability and vegetation vulnerability in Amazonia. Geophysical Research Letters 32(24).

IMF (International Monetary Fund). 2013. Energy Subsidy Reform: Lessons and Implications. Washington, DC: International Monetary Fund.

Inouye, D. W. 2008. Effects of climate change on phenology, frost damage, and floral abundance of montane wildflowers. Ecology 89(2):353-362.

IPCC (Intergovernmental Panel on Climate Change). 2000. IPCC Special Report: Emissions Scenarios. Summary for Policymakers. Geneva, Switzerland: Intergovernmental Panel on Climate Change.

IPCC. 2001. IPCC Third Assessment Report: Climate Change 2001. Geneva, Switzerland: Intergovernmental Panel on Climate Change. Arendal, Norway: GRID-Arendal.

IPCC. 2007a. Climate Change 2007: Impacts, Adaptation and Vulnerability. Contribution of Working Group II to the Fourth Assessment Report of the Intergovernmental Panel on Climate Change. M.L. Parry, O.F. Canziani, J.P. Palutikof, P.J. van der Linden and C.E. Hanson, eds. Cambridge, UK and New York: Cambridge University Press.

IPCC. 2007b. Climate Change 2007: The Physical Science Basis. Contribution of Working Group I to the Fourth Assessment Report of the Intergovernmental Panel on Climate Change. S. Solomon, D. Qin, M. Manning, Z. Chen, M. Marquis, K. B. Averyt, M. Tignor, and H. L. Miller, eds. Cambridge, UK and New York: Cambridge University Press.

IPCC. 2007c. IPCC Fourth Assessment Report: Climate Change 2007. Geneva, Switzerland: Intergovernmental Panel on Climate Change.

IPCC. 2012. Managing the Risks of Extreme Events and Disasters to Advance Climate Change Adaptation. C.B. Field, V. Barros, T.F. Stocker, D. Qin, D.J. Dokken, K.L. Ebi, M.D. Mastrandrea, K.J. Mach, G.-K. Plattner, S.K. Allen, M. Tignor, and P.M. Midgley, eds. A Special Report of Working Groups I and II of the Intergovernmental Panel on Climate Change. Cambridge, UK and New York: Cambridge University Press.

IPCC. 2013. Climate Change 2013: The Physical Science Basis. Working Group I Contribution to the Fifth Assessment Report of the Intergovernmental Panel on Climate Change. Geneva, Switzerland: World Meteorological Organization.

Isaksen, I. S. A., M. Gauss, G. Myhre, K. M. W. Anthony and C. Ruppel. 2011. Strong atmospheric chemistry feedback to climate warming from Arctic methane emissions. Global Biogeochemical Cycles 25(2).

Ivanov, V. Y., L. R. Hutyra, S. C. Wofsy, J. W. Munger, S. R. Saleska, R. C. de Oliveira and P. B. de Camargo. 2012. Root niche separation can explain avoidance of seasonal drought stress and vulnerability of overstory trees to extended drought in a mature Amazonian forest. Water Resources Research 48(12).

Jablonski, D. 2001. Lessons from the past: Evolutionary impacts of mass extinctions. Proceedings of the National Academy of Sciences of the United States of America 98:5393-5398.

Jackson, J. B. C. 2008. Ecological extinction and evolution in the brave new ocean. Proceedings of the National Academy of Sciences of the United States of America 105(Supplement 1):11458-11465.

Jackson, S. T. and C. Y. Weng. 1999. Late quaternary extinction of a tree species in eastern North America. Proceedings of the National Academy of Sciences of the United States of America 96(24):13847-13852.

Jakobsson, M., J. B. Anderson, F. O. Nitsche, J. A. Dowdeswell, R. Gyllencreutz, N. Kirchner, R. Mohammad, M. O'Regan, R. B. Alley, S. Anandakrishnan, B. Eriksson, A. Kirshner, R. Fernandez, T. Stolldorf, R. Minzoni and W. Majewski. 2011. Geological record of ice shelf break-up and grounding line retreat, Pine Island Bay, West Antarctica. Geology 39(7):691-694.

Jansen, E., J. Overpeck, K. R. Briffa, J.-C. Duplessy, F. Joos, V. Masson-Delmotte, D. Olago, B. Otto-Bliesner, W. R. Peltier, S. Rahmstorf, R. Ramesh, D. Raynaud, D. Rind, O. Solomina, R. Villalba and D. Zhang. 2007. Paleoclimate. In Climate Change 2007: The Physical Science Basis. Contribution of Working Group I to the Fourth Assessment Report of the Intergovernmental Panel on Climate Change. Solomon, S., D. Qin, M. Manning, Z. Chen, M. Marquis, K. B. Averyt, M. Tignor and H. L. Miller, eds. Cambridge, UK and New York: Cambridge University Press.

Jansen, T., A. Campbell, C. Kelly, H. Hatun and M. R. Payne. 2012. Migration and fisheries of North East Atlantic mackerel (Scomber scombrus) in autumn and winter. PLoS ONE 7(12).

Johnstone, J. A. and T. E. Dawson. 2010. Climatic context and ecological implications of summer fog decline in the coast redwood region. Proceedings of the National Academy of Sciences of the United States of America 107(10):4533-4538.

Jones, B. M., C. D. Arp, M. T. Jorgenson, K. M. Hinkel, J. A. Schmutz and P. L. Flint. 2009. Increase in the rate and uniformity of coastline erosion in Arctic Alaska. Geophysical Research Letters 36(3).

Jorgenson, M. T., Y. L. Shur and E. R. Pullman. 2006. Abrupt increase in permafrost degradation in Arctic Alaska. Geophysical Research Letters 33(2).

Joughin, I., R. B. Alley and D. M. Holland. 2012a. Ice-sheet response to oceanic forcing. Science 338(6111):1172-1176.

Joughin, I., B. E. Smith and W. Abdalati. 2010. Glaciological advances made with interferometric synthetic aperture radar. Journal of Glaciology 56(200):1026-1042.

Joughin, I., B. E. Smith, I. M. Howat, D. Floricioiu, R. B. Alley, M. Truffer and M. Fahnestock. 2012b. Seasonal to decadal scale variations in the surface velocity of Jakobshavn Isbrae, Greenland: Observation and model-based analysis. Journal of Geophysical Research-Earth Surface 117(F2).

Jung, T., M. J. Miller, T. N. Palmer, P. Towers, N. Wedi, D. Achuthavarier, J. M. Adams, E. L. Altshuler, B. A. Cash, J. L. Kinter, L. Marx, C. Stan and K. I. Hodges. 2012. High-resolution global climate simulations with the ECMWF Model in Project Athena: Experimental design, model climate, and seasonal forecast skill. Journal of Climate 25(9):3155-3172.

Jungclaus, J. H., H. Haak, M. Esch, E. Roeckner and J. Marotzke. 2006. Will Greenland melting halt the thermohaline circulation? Geophysical Research Letters 33(17).

Kahrl, F. and D. Roland-Holst. 2012. Climate Change in California. Berkeley: University of California Press.

Karnauskas, K. B., R. Seager, A. Kaplan, Y. Kushnir and M. A. Cane. 2009. Observed strengthening of the zonal sea surface temperature gradient across the Equatorial Pacific Ocean. Journal of Climate 22(16):4316-4321.

Kasperson, R. E., O. Renn, P. Slovic, H. S. Brown, J. Emel, R. Goble, J. X. Kasperson and S. Ratick. 1988. The social amplification of risk—A conceptual framework. Risk Analysis 8(2):177-187.

Kay, J. E., M. M. Holland and A. Jahn. 2011. Inter-annual to multi-decadal Arctic sea ice extent trends in a warming world. Geophysical Research Letters 38(15).

Keller, K., B. M. Bolker and D. F. Bradford. 2004. Uncertain climate thresholds and optimal economic growth. Journal of Environmental Economics and Management 48(1):723-741.

Keller, K., S.-R. Kim, J. Baehr, D. B. Bradford and M. Oppenheimer. 2007. What is the economic value of information about climate thresholds? In Integrated Assessment of Human Induced Climate Change. Schlesinger, M., eds. Cambridge, UK: Cambridge University Press.

Keller, K., G. Yohe and M. Schlesinger. 2008. Managing the risks of climate thresholds: uncertainties and information needs. Climatic Change 91(1-2):5-10.

Kelly, R., M. L. Chipman, P. E. Higuera, I. Stefanova, L. B. Brubaker and F. S. Hu. 2013. Recent burning of boreal forests exceeds fire regime limits of the past 10,000 years. Proceedings of the National Academy of Sciences 110(32).

Kennett, D. J., S. F. M. Breitenbach, V. V. Aquino, Y. Asmerom, J. Awe, J. U. L. Baldini, P. Bartlein, B. J. Culleton, C. Ebert, C. Jazwa, M. J. Macri, N. Marwan, V. Polyak, K. M. Prufer, H. E. Ridley, H. Sodemann, B. Winterhalder and G. H. Haug. 2012. Development and disintegration of Maya political systems in response to climate change. Science 338(6108):788-791.

Kessler, J. D., D. L. Valentine, M. C. Redmond, M. R. Du, E. W. Chan, S. D. Mendes, E. W. Quiroz, C. J. Villanueva, S. S. Shusta, L. M. Werra, S. A. Yvon-Lewis and T. C. Weber. 2011. A persistent oxygen anomaly reveals the fate of spilled methane in the deep Gulf of Mexico. Science 331(6015):312-315.

Khvorostyanov, D. V., P. Ciais, G. Krinner and S. A. Zimov. 2008a. Vulnerability of east Siberia's frozen carbon stores to future warming. Geophysical Research Letters 35(10).

Khvorostyanov, D. V., P. Ciais, G. Krinner, S. A. Zimov, C. Corradi and G. Guggenberger. 2008b. Vulnerability of permafrost carbon to global warming. Part II: Sensitivity of permafrost carbon stock to global warming. Tellus Series B—Chemical and Physical Meteorology 60(2):265-275.

Khvorostyanov, D. V., G. Krinner, P. Ciais, M. Heimann and S. A. Zimov. 2008c. Vulnerability of permafrost carbon to global warming. Part I: Model description and role of heat generated by organic matter decomposition. Tellus Series B—Chemical and Physical Meteorology 60(2):250-264.

Kinsella, S., T. Spencer and B. Farling. 2008. Trout in trouble: The impacts of global warming on trout in the interior West. NRDC Issue Paper:1-36.

Kleidon, A. and M. Heimann. 2000. Assessing the role of deep rooted vegetation in the climate system with model simulations: mechanism, comparison to observations and implications for Amazonian deforestation. Climate Dynamics 16(2-3):183-199.

Klein, A. M., B. E. Vaissiere, J. H. Cane, I. Steffan-Dewenter, S. A. Cunningham, C. Kremen and T. Tscharntke. 2007. Importance of pollinators in changing landscapes for world crops. Proceedings of the Royal Society B—Biological Sciences 274(1608):303-313.

Knoll, A. H., R. K. Bambach, J. L. Payne, S. Pruss and W. W. Fischer. 2007. Paleophysiology and end-Permian mass extinction. Earth and Planetary Science Letters 256:295-313.

Knowlton, N. and J. B. C. Jackson. 2008. Shifting baselines, local impacts, and global change on coral reefs. PLoS Biology 6(2):215-220.

Knutti, R., D. Masson and A. Gettelman. 2013. Climate model genealogy: Generation CMIP5 and how we got there. Geophysical Research Letters 40(6):1194-1199.

Koch, P. L. and A. D. Barnosky. 2006. Late Quaternary extinctions: State of the debate. Annual Review of Ecology Evolution and Systematics 37:215-250.

Kolmannskog, V. 2008. Climate of Displacement, Climate for Protection? Copenhagen, Denmark: Danish Institute for International Studies.

Kriegler, E., J. W. Hall, H. Held, R. Dawson and H. J. Schellnhuber. 2009. Imprecise probability assessment of tipping points in the climate system. Proceedings of the National Academy of Sciences of the United States of America 106(13):5041-5046.

Krushelnycky, P. D., L. L. Loope, T. W. Giambelluca, F. Starr, K. Starr, D. R. Drake, A. D. Taylor and R. H. Robichaux. 2013. Climate-associated population declines reverse recovery and threaten future of an iconic high-elevation plant. Global Change Biology 19(3):911-922.

Kuparinen, A., O. Savolainen and F. M. Schurr. 2010. Increased mortality can promote evolutionary adaptation of forest trees to climate change. Forest Ecology and Management 259(5):1003-1008.

Kurz, W. A., C. C. Dymond, G. Stinson, G. J. Rampley, E. T. Neilson, A. L. Carroll, T. Ebata and L. Safranyik. 2008. Mountain pine beetle and forest carbon feedback to climate change. Nature 452(7190):987-990.

Kushner, P. J., I. M. Held and T. L. Delworth. 2001. Southern Hemisphere atmospheric circulation response to global warming. Journal of Climate 14(10):2238-2249.

Kwadijk, J. C. J., M. Haasnoot, J. P. M. Mulder, M. M. C. Hoogvliet, A. B. M. Jeuken, R. A. A. van der Krogt, N. G. C. van Oostrom, H. A. Schelfhout, E. H. van Velzen, H. van Waveren and M. J. M. de Wit. 2010. Using adaptation tipping points to prepare for climate change and sea level rise: A case study in the Netherlands. Wiley Interdisciplinary Reviews-Climate Change 1(5):729-740.

Kwok, R., G. F. Cunningham, M. Wensnahan, I. Rigor, H. J. Zwally and D. Yi. 2009. Thinning and volume loss of the Arctic Ocean sea ice cover: 2003-2008. Journal of Geophysical Research-Oceans 114(C7).

L'Heureux, M. L., S. Lee and B. Lyon. 2013. Recent multidecadal strengthening of the Walker circulation across the tropical Pacific. Nature Climate Change, in press.

Lamarque, J. F., D. T. Shindell, B. Josse, P. J. Young, I. Cionni, V. Eyring, D. Bergmann, P. Cameron-Smith, W. J. Collins, R. Doherty, S. Dalsoren, G. Faluvegi, G. Folberth, S. J. Ghan, L. W. Horowitz, Y. H. Lee, I. A. MacKenzie, T. Nagashima, V. Naik, D. Plummer, M. Righi, S. T. Rumbold, M. Schulz, R. B. Skeie, D. S. Stevenson, S. Strode, K. Sudo, S. Szopa, A. Voulgarakis and G. Zeng. 2013. The Atmospheric Chemistry and Climate Model Intercomparison Project (ACCMIP): Overview and description of models, simulations and climate diagnostics. Geoscientific Model Development 6(1):179-206.

Lau, N. C., A. Leetmaa and M. J. Nath. 2008. Interactions between the responses of north American climate to El Nino-La Nina and to the secular warming trend in the Indian-Western Pacific Oceans. Journal of Climate 21(3):476-494.

Lavell, A., M. Oppenheimer, C. Diop, J. Hess, R. Lempert, J. Li, R. Muir-Wood and S. Myeong. 2012. Climate change: New dimensions in disaster risk, exposure, vulnerability, and resilience. In Managing the Risks of Extreme Events and Disasters to Advance Climate Change Adaptation. A Special Report of Working Groups I and II of the Intergovernmental Panel on Climate Change (IPCC). Field, C. B., V. Barros, T. F. Stocker, D. Qin, D. J. Dokken, K. L. Ebi, M. D. Mastrandrea, K. J. Mach, G.-K. Plattner, S. K. Allen, M. Tignor and P. M. Midgley, eds. Cambridge, UK and New York: Cambridge University Press.

Lavender, K. L., R. E. Davis and W. B. Owens. 2000. Mid-depth recirculation observed in the interior Labrador and Irminger seas by direct velocity measurements. Nature 407(6800):66-69.

Lawrence, D. M. and A. G. Slater. 2005. A projection of severe near-surface permafrost degradation during the 21st century. Geophysical Research Letters 32(24).

Lawrence, D. M., A. G. Slater, R. A. Tomas, M. M. Holland and C. Deser. 2008. Accelerated Arctic land warming and permafrost degradation during rapid sea ice loss. Geophysical Research Letters 35(11).

Laxon, S., N. Peacock and D. Smith. 2003. High interannual variability of sea ice thickness in the Arctic region. Nature 425(6961):947-950.

Lazier, J., R. Hendry, A. Clarke, I. Yashayaev and P. Rhines. 2002. Convection and restratification in the Labrador Sea, 1990-2000. Deep-Sea Research Part I—Oceanographic Research Papers 49(10):1819-1835.

Ledru, M. P., M. L. Salgado-Labouriau and M. L. Lorscheitter. 1998. Vegetation dynamics in southern and central Brazil during the last 10,000 yr BP. Review of Palaeobotany and Palynology 99(2):131-142.

Lehman, S. J. and L. D. Keigwin. 1992. Sudden changes in North-Atlantic circulation during the last deglaciation. Nature 356(6372):757-762.

Lemke, P., J. Ren, R. B. Alley, I. Allison, J. Carrasco, G. Flato, Y. Fujii, G. Kaser, P. Mote, R. H. Thomas and T. Zhang. 2007. Observations: Changes in Snow, Ice and Frozen Ground. In Climate Change 2007: The Physical Science Basis. Contribution of Working Group I to the Fourth Assessment Report of the Intergovernmental Panel on Climate Change. Solomon, S., D. Qin, M. Manning, Z. Chen, M. Marquis, K. B. Averyt, M. Tignor and H. L. Miller, eds. Cambridge, UK: Cambridge University Press.

Lenton, T. M. 2011. Early warning of climate tipping points. Nature Climate Change 1(4):201-209.

Lenton, T. M. 2013. What early warning systems are there for environmental shocks? Environmental Science & Policy 27:S60-S75.

Lenton, T. M., H. Held, E. Kriegler, J. W. Hall, W. Lucht, S. Rahmstorf and H. J. Schellnhuber. 2008. Tipping elements in the Earth's climate system. Proceedings of the National Academy of Sciences of the United States of America 105(6):1786-1793.

Levermann, A., P. U. Clark, B. Marzeion, G. A. Milne, D. Pollard, V. Radic and A. Robinson. 2013. The multimillennial sea-level commitment of global warming. Proceedings of the National Academy of Sciences of the United States of America 110(34):13745-13750.

Levis, S., G. B. Bonan, E. Kluzek, P. E. Thornton, A. Jones, W. J. Sacks and C. J. Kucharik. 2012. Interactive Crop Management in the Community Earth System Model (CESM1): Seasonal influences on land-atmosphere fluxes. Journal of Climate 25(14):4839-4859.

Li, W. H. and R. Fu. 2004. Transition of the large-scale atmospheric and land surface conditions from the dry to the wet season over Amazonia as diagnosed by the ECMWF re-analysis. Journal of Climate 17(13):2637-2651.

Li, W. H., R. Fu and R. E. Dickinson. 2006. Rainfall and its seasonality over the Amazon in the 21st century as assessed by the coupled models for the IPCC AR4. Journal of Geophysical Research—Atmospheres 111(D2).

Li, W. H., R. Fu, R. I. N. Juarez and K. Fernandes. 2008. Observed change of the standardized precipitation index, its potential cause and implications to future climate change in the Amazon region. Philosophical Transactions of the Royal Society B-Biological Sciences 363(1498):1767-1772.

Lindsay, R. W., J. Zhang, A. Schweiger, M. Steele and H. Stern. 2009. Arctic sea ice retreat in 2007 follows thinning trend. Journal of Climate 22(1):165-176.

Lipscomb, W. H., J. G. Fyke, M. Vizcaino, W. J. Sacks, J. Wolfe, M. Vertenstein, A. Craig, E. Kluzek and D. M. Lawrence. 2013. Implementation and initial evaluation of the Glimmer Community Ice Sheet Model in the Community Earth System Model. Journal of Climate 26:7352-7371.

Liu, J. P. and J. A. Curry. 2010. Accelerated warming of the Southern Ocean and its impacts on the hydrological cycle and sea ice. Proceedings of the National Academy of Sciences of the United States of America 107(34):14987-14992.

Liu, M. J. and J. Kronbak. 2010. The potential economic viability of using the Northern Sea Route (NSR) as an alternative route between Asia and Europe. Journal of Transport Geography 18(3):434-444.

Liu, W., X. Huang, J. Lin and M. He. 2012. Seawater acidification and elevated temperature affect gene expression patterns of the pearl oyster *Pinctada fucata*. PLoS ONE 7(3):e33679.

Liu, Z. Y., Y. Wang, R. Gallimore, M. Notaro and I. C. Prentice. 2006. On the cause of abrupt vegetation collapse in North Africa during the Holocene: Climate variability vs. vegetation feedback. Geophysical Research Letters 33(22).

Livina, V. N. and T. M. Lenton. 2007. A modified method for detecting incipient bifurcations in a dynamical system. Geophysical Research Letters 34(3).

Loarie, S. R., P. B. Duffy, H. Hamilton, G. P. Asner, C. B. Field and D. D. Ackerly. 2009. The velocity of climate change. Nature 462:1052-1055.

Lobell, D. B., C. B. Field, K. N. Cahill and C. Bonfils. 2006. Impacts of future climate change on California perennial crop yields: Model projections with climate and crop uncertainties. Agricultural and Forest Meteorology 141:208-218.

Lobell, D. B. and S. M. Gourdji. 2012. The influence of climate change on global crop productivity. Plant Physiology 160(4):1686-1697.

Lobell, D. B., W. Schlenker and J. Costa-Roberts. 2011. Climate trends and global crop production since 1980. Science 333(6042):616-620.

Long, M. C., K. Lindsay, S. Peacock, J. K. Moore and S. C. Doney. 2013. Twentieth-century oceanic carbon uptake and storage in CESM1(BGC). Journal of Climate 26:6775-6800.

Lowe, J. A. and J. M. Gregory. 2006. Understanding projections of sea level rise in a Hadley Centre coupled climate model. Journal of Geophysical Research-Oceans 111(C11).

Lozier, M. S. 2012. Overturning in the North Atlantic. Annual Review of Marine Science 4(4):291-315.

Lozier, M. S., V. Roussenov, M. S. C. Reed and R. G. Williams. 2010. Opposing decadal changes for the North Atlantic meridional overturning circulation. Nature Geoscience 3(10):728-734.

Lu, J., C. Deser and T. Reichler. 2009. Cause of the widening of the tropical belt since 1958. Geophysical Research Letters 36(L03803).

Lu, J., G. A. Vecchi and T. Reichler. 2007. Expansion of the Hadley cell under global warming. Geophysical Research Letters 34(6).

Lumpkin, R., K. G. Speer and K. P. Koltermann. 2008. Transport across 48 degrees N in the Atlantic Ocean. Journal of Physical Oceanography 38(4):733-752.

Lynch, A. H. and W. L. Wu. 2000. Impacts of fire and warming on ecosystem uptake in the boreal forest. Journal of Climate 13(13):2334-2338.

MacAyeal, D. R., T. A. Scambos, C. L. Hulbe and M. A. Fahnestock. 2003. Catastrophic ice-shelf break-up by an ice-shelf-fragment-capsize mechanism. Journal of Glaciology 49(164):22-36.

MacDonald, G. M., D. W. Beilman, K. V. Kremenetski, Y. W. Sheng, L. C. Smith and A. A. Velichko. 2006. Rapid early development of circumarctic peatlands and atmospheric CH_4 and CO_2 variations. Science 314(5797):285-288.

Mack, M. C., M. S. Bret-Harte, T. N. Hollingsworth, R. R. Jandt, E. A. G. Schuur, G. R. Shaver and D. L. Verbyla. 2011. Carbon loss from an unprecedented Arctic tundra wildfire. Nature 475(7357):489-492.

Manabe, S., M. J. Spelman and R. J. Stouffer. 1992. Transient responses of a coupled ocean atmosphere model to gradual changes of atmospheric CO_2 .2. Seasonal response. Journal of Climate 5(2):105-126.

Mandelbrot, B. 1963. The Variation of certain speculative prices. Journal of Business 36(4):394-419.

Manganello, J. V., K. I. Hodges, J. L. Kinter, B. A. Cash, L. Marx, T. Jung, D. Achuthavarier, J. M. Adams, E. L. Altshuler, B. H. Huang, E. K. Jin, C. Stan, P. Towers and N. Wedi. 2012. Tropical cyclone climatology in a 10-km global atmospheric GCM: Toward weather-resolving climate modeling. Journal of Climate 25(11):3867-3893.

Mantua, N. J., S. R. Hare, Y. Zhang, J. M. Wallace and R. C. Francis. 1997. A Pacific interdecadal climate oscillation with impacts on salmon production. Bulletin of the American Meteorological Society 78(6):1069-1079.

Marengo, J. A., C. A. Nobre, J. Tomasella, M. F. Cardoso and M. D. Oyama. 2008. Hydro-climatic and ecological behaviour of the drought of Amazonia in 2005. Philosophical Transactions of the Royal Society B—Biological Sciences 363(1498):1773-1778.

Maslowski, W., J. C. Kinney, M. Higgins and A. Roberts. 2012. The future of Arctic sea ice. Annual Review of Earth and Planetary Sciences 40(40):625-654.

Massonnet, F., T. Fichefet, H. Goosse, C. M. Bitz, G. Philippon-Berthier, M. M. Holland and P. Y. Barriat. 2012. Constraining projections of summer Arctic sea ice. Cryosphere 6(6):1383-1394.

Maurer, B. A. 1999. Untangling Ecological Complexity :The Macroscopic Perspective. Chicago, IL: University of Chicago Press.

Mayer, A. L. and A. H. Khalyani. 2011. Grass trumps trees with fire. Science 334(6053):188-189.

McCarthy, G., E. Frajka-Williams, W. E. Johns, M. O. Baringer, C. S. Meinen, H. L. Bryden, D. Rayner, A. Duchez, C. Roberts and S. A. Cunningham. 2012. Observed interannual variability of the Atlantic meridional overturning circulation at 26.5 degrees N. Geophysical Research Letters 39.

McElroy, M. and D. J. Baker. 2012. Climate Extremes: Recent Trends with Implications for National Security. Cambridge, MA: Harvard University Press.

McGuire, A. D., L. G. Anderson, T. R. Christensen, S. Dallimore, L. D. Guo, D. J. Hayes, M. Heimann, T. D. Lorenson, R. W. Macdonald and N. Roulet. 2009. Sensitivity of the carbon cycle in the Arctic to climate change. Ecological Monographs 79(4):523-555.

McInerney, D. and K. Keller. 2008. Economically optimal risk reduction strategies in the face of uncertain climate thresholds. Climatic Change 91(1-2):29-41.

McInerney, D., R. Lempert and K. Keller. 2012. What are robust strategies in the face of uncertain climate threshold responses? Robust climate strategies. Climatic Change 112(3-4):547-568.

McLachlan, J. S., J. J. Hellmann and M. W. Schwartz. 2007. A framework for debate of assisted migration in an era of climate change. Conservation Biology 21(2):297-302.

McLandress, C. and T. G. Shepherd. 2009. Simulated anthropogenic changes in the Brewer-Dobson Circulation, including its extension to high latitudes. Journal of Climate 22(6):1516-1540.

McLandress, C., T. G. Shepherd, J. F. Scinocca, D. A. Plummer, M. Sigmond, A. I. Jonsson and M. C. Reader. 2011. Separating the dynamical effects of climate change and ozone depletion.:Part II Southern Hemisphere troposphere. Journal of Climate 24(6):1850-1868.

McLaughlin, J. F., J. J. Hellmann, C. L. Boggs and P. R. Ehrlich. 2002. Climate change hastens population extinctions. Proceedings of the National Academy of Sciences of the United States of America 99(9):6070-6074.

McMenamin, S. K., E. A. Hadly and C. K. Wright. 2008. Climatic change and wetland desiccation cause amphibian decline in Yellowstone National Park. Proceedings of the National Academy of Sciences of the United States of America 105(44):16988-16993.

Meehl, G. A., C. Covey, T. Delworth, M. Latif, B. McAvaney, J. F. B. Mitchell, R. J. Stouffer and K. E. Taylor. 2007a. The WCRP CMIP3 multimodel dataset—A new era in climate change research. Bulletin of the American Meteorological Society 88(9): 1383-1394

Meehl, G. A., T. F. Stocker, W. D. Collins, P. Friedlingstein, A. T. Gaye, J. M. Gregory, A. Kitoh, R. Knutti, J. M. Murphy, A. Noda, S. C. B. Raper, I. G. Watterson, A. J. Weaver and Z.-C. Zhao. 2007b. Global Climate Projections. In Climate Change 2007: The Physical Science Basis. Contribution of Working Group I to the Fourth Assessment Report of the Intergovernmental Panel on Climate Change. Solomon, S., D. Qin, M. Manning, Z. Chen, M. Marquis, K. B. Averyt, M. Tignor and H. L. Miller, eds. Cambridge, UK and New York: Cambridge University Press.

Meehl, G. A. and H. Y. Teng. 2007. Multi-model changes in El Niño teleconnections over North America in a future warmer climate. Climate Dynamics 29(7-8):779-790.

Meinen, C. S., D. R. Watts and R. A. Clarke. 2000. Absolutely referenced geostrophic velocity and transport on a section across the North Atlantic Current. Deep-Sea Research Part I-Oceanographic Research Papers 47(2):309-322.

Merryfield, W. J., M. M. Holland and A. H. Monahan. 2008., 2008: Multiple equilibria and abrupt transitions in Arctic summer sea ice extent. In Arctic Sea Ice Decline: Observations, Projections, Mechanisms, and Implications. Geophysical Monograph Series, Vol. 180. DeWeaver, E. T., C. M. Bitz and L.-B. Tremblay, eds. Washington, DC: American Geophysical Union.

Metz, B., Intergovernmental Panel on Climate Change and Working Group III. 2007. Climate Change 2007: Mitigation of Climate Change: Contribution of Working Group III to the Fourth Assessment report of the Intergovernmental Panel on Climate Change. Cambridge; New York: Cambridge University Press.

Mienert, J., M. Vanneste, S. Bunz, K. Andreassen, H. Haflidason and H. P. Sejrup. 2005. Ocean warming and gas hydrate stability on the mid-Norwegian margin at the Storegga Slide. Marine and Petroleum Geology 22(1-2):233-244.

Miguel, E., S. Satyanath and E. Sergenti. 2004. Economic shocks and civil conflict: An instrumental variables approach. Journal of Political Economy 112(4):725-753.

Mikolajewicz, U., M. Vizcaino, J. Jungclaus and G. Schurgers. 2007. Effect of ice sheet interactions in anthropogenic climate change simulations. Geophysical Research Letters 34(18).

Mileti, D. 1999. Disasters by Design: A Reassessment of Natural Hazards in the United States. Washington, DC: Joseph Henry Press.

Miller, A. W., A. C. Reynolds, C. Sobrino and G. F. Riedel. 2009. Shellfish face uncertain future in high CO_2 world: Influence of acidification on oyster larvae calcification and growth in estuaries. PLoS ONE 4(5):e5661.

Miller, R. L., G. A. Schmidt and D. T. Shindell. 2006. Forced annular variations in the 20th century intergovernmental panel on climate change fourth assessment report models. Journal of Geophysical Research—Atmospheres 111(D18).

Mitrovica, J. X., M. E. Tamisiea, J. L. Davis and G. A. Milne. 2001. Recent mass balance of polar ice sheets inferred from patterns of global sea-level change. Nature 409(6823):1026-1029.

Moore, J., A. Grinsted, T. Zwinger and S. Jevrejeva. 2013. Semiempirical and process-based global sea level projections. Reviews of Geophysics 51(3).

Moritz, C. and R. Agudo. 2013. The future of species under climate change: Resilience or decline? Science 341(6145):504-508.

Moritz, C., J. L. Patton, C. J. Conroy, J. L. Parra, G. C. White and S. R. Beissinger. 2008. Impact of a century of climate change on small-mammal communities in Yosemite National Park, USA. Science 322(5899):261-264.

Moss, R. H., J. A. Edmonds, K. A. Hibbard, M. R. Manning, S. K. Rose, D. P. van Vuuren, T. R. Carter, S. Emori, M. Kainuma, T. Kram, G. A. Meehl, J. F. B. Mitchell, N. Nakicenovic, K. Riahi, S. J. Smith, R. J. Stouffer, A. M. Thomson, J. P. Weyant and T. J. Wilbanks. 2010. The next generation of scenarios for climate change research and assessment. Nature 463(7282):747-756.

Muller-Stoffels, M. and R. Wackerbauer. 2011. Regular network model for the sea ice-albedo feedback in the Arctic. Chaos 21(1).

Mumby, P. J., A. Hastings and H. J. Edwards. 2007. Thresholds and the resilience of Caribbean coral reefs. Nature 450:98-101.

Munday, P. L., D. L. Dixson, J. M. Donelsona, G. P. Jonesa, M. S. Pratchetta, G. V. Devitsinac and K. B. Døvingd. 2009. Ocean acidification impairs olfactory discrimination and homing ability of a marine fish. Proceedings of the California Academy of Sciences 106(6):1848-1852.

National Climate Assessment and Development Advisory Committee. 2013. National Climate Assessment. Washington, DC: U.S. Global Change Research Program.

Nel, P. and M. Righarts. 2008. Natural disasters and the risk of violent civil conflict. International Studies Quarterly 52(1):159-185.

Nelson, F. E., O. A. Anisimov and N. I. Shiklomanov. 2001. Subsidence risk from thawing permafrost—The threat to man-made structures across regions in the far north can be monitored. Nature 410(6831):889-890.

Nelson, F. E., O. A. Anisimov and N. I. Shiklomanov. 2002. Climate change and hazard zonation in the circum-Arctic permafrost regions. Natural Hazards 26(3):203-225.

Nepstad, D. C., I. M. Tohver, D. Ray, P. Moutinho and G. Cardinot. 2007. Mortality of large trees and lianas following experimental drought in an amazon forest. Ecology 88(9):2259-2269.

Nettles, M. and G. Ekstrom. 2010. Glacial Earthquakes in Greenland and Antarctica. Annual Review of Earth and Planetary Sciences 38(38):467-491.

New, M., M. Hulme and P. Jones. 1999. Representing twentieth-century space-time climate variability. Part I: Development of a 1961-90 mean monthly terrestrial climatology. Journal of Climate 12(3):829-856.

Newman, M., G. P. Compo and M. A. Alexander. 2003. ENSO-forced variability of the Pacific decadal oscillation. Journal of Climate 16(23):3853-3857.

Nicholls, R. J., P. P. Wong, V. R. Burkett, J. O. Codignotto, J. E. Hay, R. F. McLean, S. Ragoonaden and C. D. Woodroffe. 2007. Coastal systems and low-lying areas. In Climate Change 2007: Impacts, Adaptation and Vulnerability. Contribution of Working Group II to the Fourth Assessment Report of the Intergovernmental Panel on Climate Change. Parry, M. L., O. F. Canziani, J. P. Palutikof, P. J. v. d. Linden and C. E. Hanson, eds. Cambridge, UK: Cambridge University Press.

Nix, H. A. 1983. Climate of tropical savannas. In Tropical Savannas. Ecosystems of the World, vol. 13. F. Bourliere, ed. New York: Elsevier.

NOAA (National Oceanic and Atmospheric Administration). 2013. The National Coastal Population Report, Population Trends from 1970 to 2020 Silver Spring, MD: NOAA.

Nobre, C. A. and L. D. Borma. 2009. 'Tipping points' for the Amazon forest. Current Opinion in Environmental Sustainability 1(1):28-36.

Nordhaus, W. D. 2010. The economics of hurricanes and implications of global warming. Climate Change Economics 1(1):1-20.

Normand, S., R. E. Ricklefs, F. Skov, J. Bladt, O. Tackenberg and J. C. Svenning. 2011. Postglacial migration supplements climate in determining plant species ranges in Europe. Proceedings of the Royal Society B-Biological Sciences 278(1725):3644-3653.

North, G.R. 1984. The small ice cap instability in diffusive climate models. Journal of the Atmospheric Sciences 41(23):3390-3395.

North, G. R. 1990. Multiple solutions in energy-balance climate models. Global and Planetary Change 82(3-4):225-235.

Nowicki, S., R. A. Bindschadler, A. Abe-Ouchi, A. Aschwanden, E. Bueler, H. Choi, J. Fastook, G. Granzow, R. Greve, G. Gutowski, U. Herzfeld, C. Jackson, J. Johnson, C. Khroulev, E. Larour, A. Levermann, W. H. Lipscomb, M. A. Martin, M. Morlighem, B. R. Parizek, D. Pollard, S. F. Price, D. Ren, E. Rignot, F. Saito, T. Sato, H. Seddik, H. Seroussi, K. Takahashi, R. Walker and W. L. Wang. 2013. Insights into spatial sensitivities of ice mass response to environmental change from the SeaRISE ice sheet modeling project I: Antarctica. Journal of Geophysical Research-Earth Surface, accepted.

NRC (National Research Council). 2002. Abrupt Climate Change : Inevitable Surprises. Washington, DC: National Academy Press.

NRC. 2007. Polar Icebreakers in a Changing World: An Assessment of U.S. Needs. Washington, DC: National Academies Press.

NRC. 2010a. Adapting to the Impacts of Climate Change. Washington, DC: National Academies Press.

NRC. 2010b. Informing an Effective Response to Climate Change. Washington, DC: National Academies Press.

NRC. 2010c. Ocean Acidification: A National Strategy to Meet the Challenges of a Changing Ocean. Washington, DC: National Academies Press.

NRC. 2011a. Climate Stablilization Targets: Emissions, Concentrations, and Impacts of Decades to Millennia. Washington, DC: National Academies Press.

NRC. 2011b. Scientific Ocean Drilling: Accomplishments and Challenges. Washington, DC: National Academies Press.

NRC. 2012a. Climate and Social Stress: Implications for Security Analysis. Washington, DC. National Academies Press.

NRC. 2012b. Climate and Social Stresses: Implications for Security Analysis. Washington, DC: National Academies Press.

NRC. 2012c. Disaster Resilience: A National Imperative. Washington, DC: National Academies Press.

NRC. 2012d. Earth Science and Applications from Space: A Midterm Assessment of NASA's Implementation of the Decadal Survey. Washington, DC: National Academies Press.

NRC. 2012e. Sea-Level Rise for the Coasts of California, Oregon, and Washington: Past, Present, and Future. Washington, DC: National Academies Press.

NRC. 2013. Review of the Federal Ocean Acidification Research and Monitoring Plan. Washington, DC: National Academies Press.

Numata, I., M. A. Cochrane, D. A. Roberts, J. V. Soares, C. M. Souza and M. H. Sales. 2010. Biomass collapse and carbon emissions from forest fragmentation in the Brazilian Amazon. Journal of Geophysical Research—Biogeosciences 115(G3).

O'Loughlin, J., F. D. W. Witmer, A. M. Linke, A. Laing, A. Gettelman and J. Dudhia. 2012. Climate variability and conflict risk in East Africa, 1990-2009. Proceedings of the National Academy of Sciences of the United States of America 109(45):18344-18349.

O'Rourke, R. 2013. Coast Guard Polar Icebreaker Modernization: Background and Issues for Congress. CRS Report for Congress. Washington, DC: Congressional Research Service.

O'Leary, M. J., P. J. Hearty, W. G. Thompson, M. E. Raymo, J. X. Mitrovica and J. M. Webster. 2013. Ice sheet collapse following a prolonged period of stable sea level during the last interglacial. Nature Geoscience 6:796–800.

Oleson, K. 2012. Contrasts between urban and rural climate in CCSM4 CMIP5 climate change scenarios. Journal of Climate 25(5):1390-1412.

Oliveira, P. S. and R. J. Marquis. 2002. The Cerrados of Brazil: Ecology and Natural History of a Neotropical Savanna. New York: Columbia University Press.

Ordonez, A. 2013. Realized climatic niche of North American plant taxa lagged behind climate during the end of the Pleistocene. American Journal of Botany 100(7):1255-1265.

Overland, J. E. and M. Wang. 2013. When will the summer Arctic be nearly sea ice free? Geophysical Research Letters 40:2097-2101.

Overpeck, J. T. and J. E. Cole. 2006. Abrupt change in Earth's climate system. Annual Review of Environment and Resources 31:1-31.

Pagani, M., K. Caldeira, D. Archer and J. C. Zachos. 2006. An ancient carbon mystery. Science 314(5805):1556-1557.

Pandolfi, J. M., S. R. Connolly, D. J. Marshall and A. L. Cohen. 2011. Projecting coral reef futures under global warming and ocean acidification. Science 333:418-422.

Parizek, B. R., K. Christianson, S. Anandakrishnan, R. B. Alley, R. T. Walker, R. A. Edwards, D. S. Wolfe, G. T. Bertini, S. K. Rinehart, R. A. Bindschadler and S. M. J. Nowicki. 2013. Dynamic (In)stability of Thwaites Glacier, West Antarctica. Journal of Geophysical Research, in press.

Parmesan, C. 2006. Ecological and evolutionary responses to recent climate change. Annual Review of Ecology, Evolution and Systematics 37:637-639.

Parmesan, C. and G. Yohe. 2003. A globally coherent fingerprint of climate change across natural systems. Nature 421:37-42.

Patricola, C. M. and K. H. Cook. 2008. Atmosphere/vegetation feedbacks: A mechanism for abrupt climate change over northern Africa. Journal of Geophysical Research-Atmospheres 113(D18).

Payne, J. L. and M. E. Clapham. 2012. End-Permian mass extinction in the oceans: An ancient analog for the twenty-first century? Annual Review of Earth and Planetary Sciences 40:89-111.

Pereira, H. M., S. Ferrier, M. Walters, G. N. Geller, R. H. G. Jongman, R. J. Scholes, M. W. Bruford, N. Brummitt, S. H. M. Butchart, A. C. Cardoso, N. C. Coops, E. Dulloo, D. P. Faith, J. Freyhof, R. D. Gregory, C. Heip, R. Hoft, G. Hurtt, W. Jetz, D. S. Karp, M. A. McGeoch, D. Obura, Y. Onoda, N. Pettorelli, B. Reyers, R. Sayre, J. P. W. Scharlemann, S. N. Stuart, E. Turak, M. Walpole and M. Wegmann. 2013. Essential biodiversity variables. Science 339(6117):277-278.

Peters, D. C. P., B. T. Bestelmeyer, A. K. Knapp, J. E. Herrick, H. C. Monger and K. M. Havstad. 2009. Approaches to predicting broad-scale regime shifts using changing pattern-process relationships across scales. In Real World Ecology: Large-scale and Long-Term Case Studies and Methods. Miao, S. L., S. Carstenn and M. K. Nungesser, eds. New York: Springer.

Petit, J. R., J. Jouzel, D. Raynaud, N. I. Barkov, J. M. Barnola, I. Basile, M. Bender, J. Chappellaz, M. Davis, G. Delaygue, M. Delmotte, V. M. Kotlyakov, M. Legrand, V. Y. Lipenkov, C. Lorius, L. Pepin, C. Ritz, E. Saltzman and M. Stievenard. 1999. Climate and atmospheric history of the past 420,000 years from the Vostok ice core, Antarctica. Nature 399(6735):429-436.

Pfeffer, W. T., J. T. Harper and S. O'Neel. 2008. Kinematic constraints on glacier contributions to 21st-century sea-level rise. Science 321:1340-1343.

Phillips, O. L., L. E. O. C. Aragao, S. L. Lewis, J. B. Fisher, J. Lloyd, G. Lopez-Gonzalez, Y. Malhi, A. Monteagudo, J. Peacock, C. A. Quesada, G. van der Heijden, S. Almeida, I. Amaral, L. Arroyo, G. Aymard, T. R. Baker, O. Banki, L. Blanc, D. Bonal, P. Brando, J. Chave, A. C. A. de Oliveira, N. D. Cardozo, C. I. Czimczik, T. R. Feldpausch, M. A. Freitas, E. Gloor, N. Higuchi, E. Jimenez, G. Lloyd, P. Meir, C. Mendoza, A. Morel, D. A. Neill, D. Nepstad, S. Patino, M. C. Penuela, A. Prieto, F. Ramirez, M. Schwarz, J. Silva, M. Silveira, A. S. Thomas, H. ter Steege, J. Stropp, R. Vasquez, P. Zelazowski, E. A. Davila, S. Andelman, A. Andrade, K. J. Chao, T. Erwin, A. Di Fiore, E. Honorio, H. Keeling, T. J. Killeen, W. F. Laurance, A. P. Cruz, N. C. A. Pitman, P. N. Vargas, H. Ramirez-Angulo, A. Rudas, R. Salamao, N. Silva, J. Terborgh and A. Torres-Lezama. 2009. Drought sensitivity of the Amazon rainforest. Science 323(5919):1344-1347.

Pimm, S. L. 2001. The World According to Pimm: A Scientist Audits the Earth. New York: McGraw-Hill.

Pimm, S. L. 2009. Climate Disruption and Biodiversity. Current Biology 19(14):R595-R601.

Pimm, S. L. and T. M. Brooks. 1997. The sixth extinction: How large, where, and when? In Nature and Human Society: The Quest for a Sustainable World. Raven, P. H. and T. Williams, eds. Washington, DC: National Academy Press.

Pimm, S. L., G. J. Russell, J. L. Gittleman and T. M. Brooks. 1995. The future of biodiversity. Science 269(5222):347-350.

Pinsky, M. L. and M. Fogarty. 2012. Lagged social-ecological responses to climate and range shifts in fisheries. Climatic Change 115(3-4):883-891.

Plattner, T. 2005. Modelling public risk evaluation of natural hazards: A conceptual approach. Natural Hazards and Earth System Sciences 5(3):357-366.

Poloczanska, E. S., C. J. Brown, W. J. Sydeman, W. Kiessling, D. S. Schoeman, P. J. Moore, K. Brander, J. F. Bruno, L. B. Buckley, M. T. Burrows, C. M. Duarte, B. S. Halpern, J. Holding, C. V. Kappel, M. I. O'Connor, J. M. Pandolfi, C. Parmesan, F. Schwing, S. A. Thompson and A. J. Richardson. 2013. Global imprint of climate change on marine life. Nature Climate Change doi:10.1038/nclimate1958.

Polvani, L. M., M. Previdi and C. Deser. 2011. Large cancellation, due to ozone recovery, of future Southern Hemisphere atmospheric circulation trends. Geophysical Research Letters 38(4).

Portier, C. J., K. T. Tart, S. R. Carter, C. H. Dilworth, A. E. Grambsch, J. Gohlke, J. Hess, S. N. Howard, G. Luber, J. T. Lutz, T. Maslak, N. Prudent, M. Radtke, J. P. Rosenthal, T. Rowles, P. A. Sandifer, J. Scheraga, P. J. Schramm, D. Strickman, J. M. Trtanj and P.-Y. Whung. 2010. A Human Health Perspective on Climate Change: A Report Outlining the Research Needs on the Human Health Effects of Climate Change. Research Triangle Park, NC: Environmental Health Perspectives/National Institute of Environmental Health Sciences.

Powell, A. M. and J. J. Xu. 2012. The 1977 Global regime shift: A discussion of its dynamics and impacts in the Eastern Pacific ecosystem. Atmosphere-Ocean 50(4):421-436.

Power, S. B. and I. N. Smith. 2007. Weakening of the Walker Circulation and apparent dominance of El Nino both reach record levels, but has ENSO really changed? Geophysical Research Letters 34(18).

Previdi, M. and B. G. Liepert. 2007. Annular modes and Hadley cell expansion under global warming. Geophysical Research Letters 34(22).

Pritchard, H. D., S. B. Luthcke and A. H. Fleming. 2010. Understanding ice-sheet mass balance: progress in satellite altimetry and gravimetry. Journal of Glaciology 56(200):1151-1161.

Prowse, T., K. Alfredsen, S. Beltaos, B. R. Bonsal, W. B. Bowden, C. R. Duguay, A. Korhola, J. McNamara, W. F. Vincent, V. Vuglinsky, K. M. W. Anthony and G. A. Weyhenmeyer. 2011. Effects of changes in Arctic lake and river ice. Ambio 40:63-74.

Prowse, T. D., C. Furgal, R. Chouinard, H. Melling, D. Milburn and S. L. Smith. 2009. Implications of climate change for economic development in northern Canada: Energy, resource, and transportation sectors. Ambio 38(5):272-281.

Pueyo, S., P. M. D. Graca, R. I. Barbosa, R. Cots, E. Cardona and P. M. Fearnside. 2010. Testing for criticality in ecosystem dynamics: the case of Amazonian rainforest and savanna fire. Ecology Letters 13(7):793-802.

Rahmstorf, S. 1996. On the freshwater forcing and transport of the Atlantic thermohaline circulation. Climate Dynamics 12(12):799-811.

Rahmstorf, S. 1999. Rapid transitions of the thermohaline circulation: A modelling perspective. In Reconstructing Ocean History: A Window into the Future. Abrantes, F. and A. Mix, eds. New York: Kluwer Academic/Plenum Publishers.

Rahmstorf, S., M. Crucifix, A. Ganopolski, H. Goosse, I. Kamenkovich, R. Knutti, G. Lohmann, R. Marsh, L. A. Mysak, Z. M. Wang and A. J. Weaver. 2005. Thermohaline circulation hysteresis: A model intercomparison. Geophysical Research Letters 32(23).

Randerson, J. T., H. Liu, M. G. Flanner, S. D. Chambers, Y. Jin, P. G. Hess, G. Pfister, M. C. Mack, K. K. Treseder, L. R. Welp, F. S. Chapin, J. W. Harden, M. L. Goulden, E. Lyons, J. C. Neff, E. A. G. Schuur and C. S. Zender. 2006. The impact of boreal forest fire on climate warming. Science 314(5802):1130-1132.

Rasmusson, E. M. and T. H. Carpenter. 1982. Variations in tropical sea-surface temperature and surface wind fields associated with the Southern Oscillation El-Nino. Monthly Weather Review 110(5):354-384.

Raup, D. M. and J. J. Sepkoski. 1982. Mass extinctions in the marine fossil record. Science 215(4539):1501-1503.

Rayner, D., J. J. M. Hirschi, T. Kanzow, W. E. Johns, P. G. Wright, E. Frajka-Williams, H. L. Bryden, C. S. Meinen, M. O. Baringer, J. Marotzke, L. M. Beal and S. A. Cunningham. 2011. Monitoring the Atlantic meridional overturning circulation. Deep-Sea Research Part II—Topical Studies in Oceanography 58(17-18):1744-1753.

Reeburgh, W. S. 2007. Oceanic methane biogeochemistry. Chemical Reviews 107(2):486-513.

Rehder, G., P. W. Brewer, E. T. Peltzer and G. Friederich. 2002. Enhanced lifetime of methane bubble streams within the deep ocean. Geophysical Research Letters 29(15).

Reid, W. V., H. A. Mooney, A. Cropper, D. Capistrano, S. R. Carpenter, K. Chopra, P. Dasgupta, T. Dietz, A. K. Duraiappah, R. Hassan, R. Kasperson, R. Leemans, R. M. May, A. J. McMichael, P. Pingali, C. Samper, R. Scholes, R. T. Watson, A. H. Zakri, Z. Shidong, N. J. Ash, E. Bennett, P. Kumar, M. J. Lee, C. Raudsepp-Hearne, H. Simons, J. Thonell and M. B. Zurek. 2005. Ecosystems and Human Well-Being: Synthesis: A Report of the Millennium Ecosystem Assessment. Washington, DC: Island Press.

Rhines, A. and P. Huybers. 2013. Frequent summer temperature extremes reflect changes in the mean, not the variance. Proceedings of the National Academy of Sciences of the United States of America 110(7):E546.

Ricciardi, A. and D. Simberloff. 2009. Assisted colonization is not a viable conservation strategy. Trends in Ecological Evaluation 24(5):248-253.

Ricke, K. L., J. C. Orr, K. Schneider and K. Caldeira. 2013. Risks to coral reefs from ocean carbonate chemistry changes in recent earth system model projections. Environmental Research Letters 8(3):34003-34008.

Ridley, J., J. M. Gregory, P. Huybrechts and J. Lowe. 2010. Thresholds for irreversible decline of the Greenland ice sheet. Climate Dynamics 35(6):1065-1073.

Ridley, J., J. Lowe and D. Simonin. 2008. The demise of Arctic sea ice during stabilisation at high greenhouse gas concentrations. Climate Dynamics 30(4):333-341.

Ridley, J. K., J. A. Lowe and H. T. Hewitt. 2012. How reversible is sea ice loss? Cryosphere 6(1):193-198.

Riedel, M., G. D. Spence, N. R. Chapman and R. D. Hyndman. 2002. Seismic investigations of a vent field associated with gas hydrates, offshore Vancouver Island. Journal of Geophysical Research-Solid Earth 107(B9).

Risi, C., S. Bony, F. Vimeux, C. Frankenberg, D. Noone and J. Worden. 2010. Understanding the Sahelian water budget through the isotopic composition of water vapor and precipitation. Journal of Geophysical Research-Atmospheres 115(D24).

Robertson, D. S., M. C. McKenna, O. B. Toon, S. Hope and J. A. Lillegraven. 2004. Survival in the first hours of the Cenozoic. Geological Society of America Bulletin 116:760-768.

Robine, J. M., S. L. K. Cheung, S. Le Roy, H. Van Oyen, C. Griffiths, J. P. Michel and F. R. Herrmann. 2008. Death toll exceeded 70,000 in Europe during the summer of 2003. Comptes Rendus Biologies 331(2):171-175.

Roosevelt, A. C., M. L. daCosta, C. L. Machado, M. Michab, N. Mercier, H. Valladas, J. Feathers, W. Barnett, M. I. daSilveira, A. Henderson, J. Sliva, B. Chernoff, D. S. Reese, J. A. Holman, N. Toth and K. Schick. 1996. Paleoindian cave dwellers in the Amazon: The peopling of the Americas. Science 272(5260):373-384.

Root, T. L., J. T. Price, K. R. Hall, S. H. Schneider, C. Rosenzweig and J. A. Pounds. 2003. Fingerprints of global warming on wild animals and plants. Nature 421:57-60.

Rosenzweig, C. and M. L. Parry. 1994. Potential impact of climate-change on world food-supply. Nature 367(6459):133-138.

Saatchi, S., S. Asefi-Najafabady, Y. Malhi, L. E. O. C. Aragao, L. O. Anderson, R. B. Myneni and R. Nemani. 2013. Persistent effects of a severe drought on Amazonian forest canopy. Proceedings of the National Academy of Sciences of the United States of America 110(2):565-570.

Salati, E., A. Dallolio, E. Matsui and J. R. Gat. 1979. Recycling of water in the Amazon Basin—Isotopic study. Water Resources Research 15(5):1250-1258.

Salati, E. and C. A. Nobre. 1991. Possible climatic impacts of tropical deforestation. Climatic Change 19(1-2):177-196.

Saleska, S. R., K. Didan, A. R. Huete and H. R. da Rocha. 2007. Amazon forests green-up during 2005 drought. Science 318(5850):612.

Saleska, S. R., S. D. Miller, D. M. Matross, M. L. Goulden, S. C. Wofsy, H. R. da Rocha, P. B. de Camargo, P. Crill, B. C. Daube, H. C. de Freitas, L. Hutyra, M. Keller, V. Kirchhoff, M. Menton, J. W. Munger, E. H. Pyle, A. H. Rice and H. Silva. 2003. Carbon in Amazon forests: Unexpected seasonal fluxes and disturbance-induced losses. Science 302(5650):1554-1557.

Sandel, B., L. Arge, B. Dalsgaard, R. G. Davies, K. J. Gaston, W. J. Sutherland and J. C. Svenning. 2011. The influence of late quaternary climate-change velocity on species endemism. Science 334(6056):660-664.

Sax, D. F., K. F. Smith and A. R. Thompson. 2009. Managed relocation: A nuanced evaluation is needed. Trends in Ecology & Evolution 24(9):472-473.

Scheffer, M. 2010. Foreseeing tipping points. Nature 467:411-412.

Scheffer, M., J. Bascompte, W. A. Brock, V. Brovkin, S. R. Carpenter, V. Dakos, H. Held, E. H. van Nes, M. Rietkerk and G. Sugihara. 2009. Early-warning signals for critical transitions. Nature 461(7260):53-59.

Scheffer, M., V. Brovkin and P. M. Cox. 2006. Positive feedback between global warming and atmospheric CO_2 concentration inferred from past climate change. Geophysical Research Letters 33(10).

Scheffer, M., S. Carpenter, J. A. Foley, C. Folke and B. Walker. 2001. Catastrophic shifts in ecosystems. Nature 413:591-596.

Scheffer, M., S. R. Carpenter, T. M. Lenton, J. Bascompte, W. Brock, V. Dakos, J. van de Koppel, I. A. van de Leemput, S. A. Levin, E. H. van Nes, M. Pascual and J. Vandermeer. 2012a. Anticipating critical transitions. Science 338(6105):344-348.

Scheffer, M., M. Hirota, M. Holmgren, E. H. V. Nesa and F. S. C. III. 2012b. Thresholds for boreal biome transitions. Proceedings of the National Academy of Sciences of the United States of America 109(52):21384-21389.

Schiermeier, Q. 2013. Oceans under surveillance: Three projects seek to track changes in Atlantic overturning circulation currents. Nature 497(7448).

Schlesinger, W. H., J. F. Reynolds, G. L. Cunningham, L. F. Huenneke, W. M. Jarrell, R. A. Virginia and W. G. Whitford. 1990. Biological feedbacks in global desertification. Science 247(4946):1043-1048.

Schmidt, G. A. and D. T. Shindell. 2003. Atmospheric composition, radiative forcing, and climate change as a consequence of a massive methane release from gas hydrates. Paleoceanography 18(1).

Schneider, S. H. 1997. Integrated assessment modeling of global climate change: transparent rational tool for policy making and opaque screen hiding value-laden assumptions? Environmental Modeling and Assessment 2:229-249.

Schoof, C. 2007. Ice sheet grounding line dynamics: Steady states, stability, and hysteresis. Journal of Geophysical Research-Earth Surface 112(F3).

Schott, F. A., J. Fischer, M. Dengler and R. Zantopp. 2006. Variability of the deep western boundary current east of the Grand Banks. Geophysical Research Letters 33(21).

Schulte, P., L. Alegret, I. Arenillas, J. A. Arz, P. J. Barton, P. R. Bown, T. J. Bralower, G. L. Christeson, P. Claeys, C. S. Cockell, G. S. Collins, A. Deutsch, T. J. Goldin, K. Goto, J. M. Grajales-Nishimura, R. A. F. Grieve, S. P. S. Gulick, K. R. Johnson, W. Kiessling, C. Koeberl, D. A. Kring, K. G. MacLeod, T. Matsui, J. Melosh, A. Montanari, J. V. Morgan, C. R. Neal, D. J. Nichols, R. D. Norris, E. Pierazzo, G. Ravizza, M. Rebolledo-Vieyra, W. U. Reimold, E. Robin, T. Salge, R. P. Speijer, A. R. Sweet, J. Urrutia-Fucugauchi, V. Vajda, M. T. Whalen and P. S. Willumsen. 2010. The Chicxulub Asteroid impact and mass extinction at the Cretaceous-Paleogene boundary. Science 327(5970):1214-1218.

Schuur, E. A. G., B. W. Abbott, W. B. Bowden, V. Brovkin, P. Camill, J. G. Canadell, J. P. Chanton, F. S. C. III, T. R. Christensen, P. Ciais, B. T. Crosby, C. I. Czimczik, G. Grosse, J. Harden, D. J. Hayes, G. Hugelius, J. D. Jastrow, J. B. Jones, T. Kleinen, C. D. Koven, G. Krinner, P. Kuhry, D. M. Lawrence, A. D. McGuire, S. M. Natali, J. A. O'Donnell, C. L. Ping, W. J. Riley, A. Rinke, V. E. Romanovsky, A. B. K. Sannel, C. Schädel, K. Schaefer, J. Sky, Z. M. Subin, C. Tarnocai, M. R. Turetsky, M. P. Waldrop, K. M. W. Anthony, K. P. Wickland, C. J. Wilson and S. A. Zimov. 2013. Expert assessment of vulnerability of permafrost carbon to climate change. Climatic Change 119(2):359-374.

Schuur, E. A. G., J. Bockheim, J. G. Canadell, E. Euskirchen, C. B. Field, S. V. Goryachkin, S. Hagemann, P. Kuhry, P. M. Lafleur, H. Lee, G. Mazhitova, F. E. Nelson, A. Rinke, V. E. Romanovsky, N. Shiklomanov, C. Tarnocai, S. Venevsky, J. G. Vogel and S. A. Zimov. 2008. Vulnerability of permafrost carbon to climate change: Implications for the global carbon cycle. Bioscience 58(8):701-714.

Schwartz, M. W., J. J. Hellmann, J. M. McLachlan, D. F. Sax, J. O. Borevitz, J. Brennan, A. E. Camacho, G. Ceballos, J. R. Clark, H. Doremus, R. Early, J. R. Etterson, D. Fielder, J. L. Gill, P. Gonzalez, N. Green, L. Hannah, D. W. Jamieson, D. Javeline, B. A. Minteer, J. Odenbaugh, S. Polasky, D. M. Richardson, T. L. Root, H. D. Safford, O. Sala, S. H. Schneider, A. R. Thompson, J. W. Williams, M. Vellend, P. Vitt and S. Zellmer. 2012. Managed relocation: Integrating the scientific, regulatory, and ethical challenges. Bioscience 62(8):732-743.

Schwartz, P. and D. Randall. 2003. An Abrupt Climate Change Scenario and Its Implications for United States National Security. Emeryville, CA: Global Business Network.

Screen, J. A. and I. Simmonds. 2010a. The central role of diminishing sea ice in recent Arctic temperature amplification. Nature 464(7293):1334-1337.

Screen, J. A. and I. Simmonds. 2010b. Increasing fall-winter energy loss from the Arctic Ocean and its role in Arctic temperature amplification. Geophysical Research Letters 37(16).

Screen, J. A., I. Simmonds, C. Deser and R. Tomas. 2012. The atmospheric response to three decades of observed Arctic sea ice loss. Journal of Climate, In press.

Seager, R. and N. Naik. 2012. A mechanisms-based approach to detecting recent anthropogenic hydroclimate change. Journal of Climate 25(1):236-261.

Seager, R. and G. A. Vecchi. 2010. Greenhouse warming and the 21st century hydroclimate of southwestern North America. Proceedings of the National Academy of Sciences of the United States of America 107(50):21277-21282.

Seibel, B. A. and J. J. Childress. 2013. The real limits to marine life: a further critique of the Respiration Index. Biogeosciences 10:2815-2819.

Seidel, D. J., Q. Fu, W. J. Randel and T. J. Reichler. 2008. Widening of the tropical belt in a changing climate. Nature Geoscience 1(1):21-24.

Seierstad, I. A. and J. Bader. 2009. Impact of a projected future Arctic Sea Ice reduction on extratropical storminess and the NAO. Climate Dynamics 33(7-8):937-943.

Sella, G. F., S. Stein, T. H. Dixon, M. Craymer, T. S. James, S. Mazzotti and R. K. Dokka. 2007. Observation of glacial isostatic adjustment in "stable" North America with GPS. Geophysical Research Letters 34(2).

Seneviratne, S. I., N. Nicholls, D. Easterling, C. M. Goodess, S. Kanae, J. Kossin, Y. Luo, J. Marengo, K. McInnes, M. Rahimi, M. Reichstein, A. Sorteberg, C. Vera and X. Zhang. 2012. Changes in climate extremes and their impacts on the natural physical environment. In Managing the Risks of Extreme Events and Disasters to Advance Climate Change Adaptation. A Special Report of Working Groups I and II of the Intergovernmental Panel on Climate Change (IPCC). Field, C. B., V. Barros, T. F. Stocker, D. Qin, D. J. Dokken, K. L. Ebi, M. D. Mastrandrea, K. J. Mach, G.-K. Plattner, S. K. Allen, M. Tignor and P. M. Midgley, eds. Cambridge, UK and New York: Cambridge University Press.

Seth, A., S. A. Rauscher, M. Rojas, A. Giannini and S. J. Camargo. 2011. Enhanced spring convective barrier for monsoons in a warmer world? Climatic Change 104(2):403-414.

Severinghaus, J. and E. Brook. 1999. Abrupt climate change at the end of the last glacial period inferred from trapped air in polar ice. Science 286:930-934.

Shakhova, N., I. Semiletov, I. Leifer, A. Salyuk, P. Rekant and D. Kosmach. 2010a. Geochemical and geophysical evidence of methane release over the East Siberian Arctic Shelf. Journal of Geophysical Research-Oceans 115(C8).

Shakhova, N., I. Semiletov, A. Salyuk, V. Yusupov, D. Kosmach and O. Gustafsson. 2010b. Extensive methane venting to the atmosphere from sediments of the East Siberian Arctic Shelf. Science 327(5970):1246-1250.

Shaw, R. G. and J. R. Etterson. 2012. Rapid climate change and the rate of adaptation: insight from experimental quantitative genetics. New Phytologist 195(4):752-765.

Shen, C., W. C. Wang, Z. Hao and W. Gong. 2007. Exceptional drought events over eastern China during the last five centuries. Climatic Change 85(3-4):453-471.

Shepard, G. H., T. Levi, E. G. Neves, C. A. Peres and D. W. Yu. 2012. Hunting in ancient and modern Amazonia: Rethinking sustainability. American Anthropologist 114(4):652-667.

Shepherd, A., E. R. Ivins, A. Geruo, V. R. Barletta, M. J. Bentley, S. Bettadpur, K. H. Briggs, D. H. Bromwich, R. Forsberg, N. Galin, M. Horwath, S. Jacobs, I. Joughin, M. A. King, J. T. M. Lenaerts, J. L. Li, S. R. M. Ligtenberg, A. Luckman, S. B. Luthcke, M. McMillan, R. Meister, G. Milne, J. Mouginot, A. Muir, J. P. Nicolas, J. Paden, A. J. Payne, H. Pritchard, E. Rignot, H. Rott, L. S. Sorensen, T. A. Scambos, B. Scheuchl, E. J. O. Schrama, B. Smith, A. V. Sundal, J. H. van Angelen, W. J. van de Berg, M. R. van den Broeke, D. G. Vaughan, I. Velicogna, J. Wahr, P. L. Whitehouse, D. J. Wingham, D. H. Yi, D. Young and H. J. Zwally. 2012. A reconciled estimate of ice-sheet mass balance. Science 338(6111):1183-1189.

Shepherd, T. G. and C. McLandress. 2011. A robust mechanism for strengthening of the Brewer-Dobson Circulation in response to climate change: critical-layer control of subtropical wave breaking. Journal of the Atmospheric Sciences 68(4):784-797.

Shiklomanov, N. I. and D. A. Streletskiy. 2013. Effect of Climate Change on Siberian Infrastructure. In Regional Environmental Changes in Siberia and Their Global Consequences. Groisman, P. Y. and G. Gutman, eds. Berlin: Springer Environmental Science and Engineering.

Shindell, D. T., O. Pechony, A. Voulgarakis, G. Faluvegi, L. Nazarenko, J. F. Lamarque, K. Bowman, G. Milly, B. Kovari, R. Ruedy and G. A. Schmidt. 2013. Interactive ozone and methane chemistry in GISS-E2 historical and future climate simulations. Atmospheric Chemistry and Physics 13(5):2653-2689.

Siegenthaler, U., T. F. Stocker, E. Monnin, D. Luthi, J. Schwander, B. Stauffer, D. Raynaud, J. M. Barnola, H. Fischer, V. Masson-Delmotte and J. Jouzel. 2005. Stable carbon cycle-climate relationship during the late Pleistocene. Science 310(5752):1313-1317.

Sinervo, B., F. Méndez-de-la-Cruz, D. B. Miles, B. Heulin, E. Bastiaans, M. V. n.-S. Cruz, R. Lara-Resendiz, N. Martínez-Méndez, M. L. a. Calderón-Espinosa, R. N. Meza-Lázaro, H. c. Gadsden, L. J. Avila, M. Morando, I. J. D. I. Riva, P. V. Sepulveda, C. F. D. Rocha, N. Ibargüengoytía, C. s. A. Puntriano, M. Massot, V. Lepetz, T. A. Oksanen, D. G. Chapple, A. M. Bauer, W. R. Branch, J. Clobert and J. W. Sites. 2010. Erosion of lizard diversity by climate change and altered thermal niches. Science 328(5980):894-899.

Skre, O., R. Baxter, R. M. M. Crawford, T. V. Callaghan and A. Fedorkov. 2002. How will the tundra-taiga interface respond to climate change? Ambio 37-46.

Smith, L. C. 2010. The World in 2050: Four Forces Shaping Civilization's Northern Future. New York: Dutton.

Smith, L. C., D. W. Beilman, K. V. Kremenetski, Y. W. Sheng, G. M. MacDonald, R. B. Lammers, A. I. Shiklomanov and E. D. Lapshina. 2012. Influence of permafrost on water storage in West Siberian peatlands revealed from a new database of soil properties. Permafrost and Periglacial Processes 23(1):69-79.

Smith, L. C., G. M. MacDonald, A. A. Velichko, D. W. Beilman, O. K. Borisova, K. E. Frey, K. V. Kremenetski and Y. Sheng. 2004. Siberian peatlands a net carbon sink and global methane source since the early Holocene. Science 303(5656):353-356.

Smith, L. C., Y. Sheng, G. M. MacDonald and L. D. Hinzman. 2005. Disappearing Arctic lakes. Science 308(5727):1429-1429.

Smith, L. C., Y. W. Sheng and G. M. MacDonald. 2007. A first pan-Arctic assessment of the influence of glaciation, permafrost, topography and peatlands on Northern Hemisphere lake distribution. Permafrost and Periglacial Processes 18(2):201-208.

Smith, L. C. and S. R. Stephenson. 2013. New Trans-Arctic shipping routes navigable by midcentury. Proceedings of the National Academy of Sciences of the United States of America 110(13):E1191-E1195.

Sokolov, A. P., C. A. Schlosser, S. Dutkiewicz, S. Paltsev, D. W. Kicklighter, H. D. Jacoby, R. G. Prinn, C. E. Forest, J. M. Reilly, C. Wang, B. S. Felzer, M. C. Sarofim, J. Scott, P. H. Stone, J. M. Melillo and J. B. Cohen. 2005. MIT Integrated Global System Model (IGSM) Version 2: Model Description and Baseline Evaluation. Available at http://dspace.mit.edu/handle/1721.1/29789; accessed June 14, 2013.

Solomon, S., G.-K. Plattner, R. Knutti and P. Friedlingstein. 2009. Irreversible climate change due to carbon dioxide emissions, PNAS, 106, doi:10.1073/pnas.0812721106. Proceedings of the National Academy of Sciences 106(6):1704-1709.

Solterra Solutions. 2012. Determining the Impact of Climate Change on Insurance Risk and the Global Community. Phase 1: Key Climate Indicators. Schaumburg, IL: Society of Actuaries.

Son, S.-W., E. P. Gerber, J. Perlwitz, L. M. Polvani, N. P. Gillett, K.-H. Seo, V. Eyring, T. G. Shepherd, D. Waugh, H. Akiyoshi, J. Austin, A. Baumgaertner, S. Bekki, P. Braesicke, C. Brühl, N. Butchart, M. P. Chipperfield, D. Cugnet, M. Dameri, S. Dhomse, S. Frith, H. Garn, R. Garcia, S. C. Hardiman, P. Jöckel, J. F. Lamarque, E. Mancini, M. Marchand, M. Michou, T. Nakamura, O. Morgenstern, G. Pitari, D. A. Plummer, J. Pyle, E. Rozanov, J. F. Scinocca, K. Shibata, D. Smale, H. Teyssèdre, W. Tian and Y. Yamashita. 2010. Impact of stratospheric ozone on Southern Hemisphere circulation change: A multimodel assessment. Journal of Geophysical Research-Solid Earth 115:D00M07.

Staver, A. C., S. Archibald and S. A. Levin. 2011. The global extent and determinants of savanna and forest as alternative biome states. Science 334(6053):230-232.

Steinacher, M., F. Joos, T. L. Frolicher, L. Bopp, P. Cadule, V. Cocco, S. C. Doney, M. Gehlen, K. Lindsay, J. K. Moore, B. Schneider and J. Segschneider. 2010. Projected 21st century decrease in marine productivity: A multi-model analysis. Biogeosciences 7(3):979-1005.

Stephens, P. A., W. J. Sutherland and R. P. Freckleton. 1999. What is the Allee effect? Oikos 87(1):185-190.

Stephenson, S. R., L. C. Smith and J. A. Agnew. 2011. Divergent long-term trajectories of human access to the Arctic. Nature Climate Change 1(3):156-160.

Stephenson, S. R., L. C. Smith, L. W. Brigham and J. A. Agnew. 2013. Projected 21st-century changes to Arctic marine access. Climatic Change 118(3-4):885-899.

Stewart, R. 2010. Desertification in the Sahel. In Environmental Science in the 21st Century: A New Online Environmental Science Book for College Students, R. Stewart, ed. Available at http://oceanworld.tamu.edu/resources/environment-book/desertificationinsahel.html; accessed September 3, 2013.

Stommel, H. 1961. Thermohaline convection with 2 stable regimes of flow. Tellus 13(2):224-230.

Stramma, L., S. Schmidtko, L. A. Levin and G. C. Johnson. 2010. Ocean oxygen minima expansions and their biological impacts. Deep-Sea Research Part I—Oceanographic Research Papers 57(4):587-595.

Strauss, B., C. Tebaldi, S. Kulp, S. Cutter, C. Emrich, D. Rizza and D. Yawitz. 2013. Florida and the Surging Sea: A Vulnerability Assessment With Projections for Sea Level Rise and Coastal Flood Risk. Climate Central, Princeton, NJ.

Strauss, B. H., R. Ziemlinski, J. L. Weiss and J. T. Overpeck. 2012. Tidally adjusted estimates of topographic vulnerability to sea level rise and flooding for the contiguous United States. Environmental Research Letters 7(1).

Streletskiy, D. A., N. I. Shiklomanov and F. E. Nelson. 2012. Permafrost, Infrastructure, and Climate Change: A GIS-Based Landscape Approach to Geotechnical Modeling. Arctic Antarctic and Alpine Research 44(3):368-380.

Stroeve, J., M. M. Holland, W. Meier, T. Scambos and M. Serreze. 2007. Arctic sea ice decline: Faster than forecast. Geophysical Research Letters 34(9).

Stroeve, J., M. Serreze, S. Drobot, S. Gearheard, M. Holland, J. Maslanik, W. Meier and T. Scambos. 2008. Arctic sea ice extent plummets in 2007. Eos, Transactions American Geophysical Union 89(2):13-14.

Stroeve, J. C., V. Kattsov, A. Barrett, M. Serreze, T. Pavlova, M. Holland and W. N. Meier. 2012a. Trends in Arctic sea ice extent from CMIP5, CMIP3 and observations. Geophysical Research Letters 39(16).

Stroeve, J. C., M. C. Serreze, M. M. Holland, J. E. Kay, J. Malanik and A. P. Barrett. 2012b. The Arctic's rapidly shrinking sea ice cover: a research synthesis. Climatic Change 110(3-4):1005-1027.

Sturm, C., Q. Zhang and D. Noone. 2010. An introduction to stable water isotopes in climate models: benefits of forward proxy modelling for paleoclimatology. Climate of the Past 6(1):115-129.

Stynes, D. J. 2011. Economic benefits to local communities from national park visitation and payroll, 2010. Natural Resource Report NPS/NRSS/EQD/NRR—2011/481: http://www.nature.nps.gov/socialscience/docs/NPSSystemEstimates2010.pdf.

Sullivan, G. R., F. Bowman, L. P. F. Jr., P. G. G. II, P. J. Kern, T. J. Lopez, D. L. Pilling, J. W. Prueher, R. H. Truly, C. F. Wald and A. C. Zinni. 2007. National Security and the Threat of Climate Change. Alexandria, VA: CNA Corporation.

Sun, Y., M. M. Joachimski, P. B. Wignall, C. Yan, Y. Chen, H. Jiang, L. Wang and X. Lai. 2012. Lethally hot temperatures during the early Triassic greenhouse. Science 338:366-370.

Swift, T. L. and S. J. Hannon. 2010. Critical thresholds associated with habitat loss: A review of the concepts, evidence, and applications. Biological Reviews 85(1):35-53.

Tarnocai, C., J. G. Canadell, E. A. G. Schuur, P. Kuhry, G. Mazhitova and S. Zimov. 2009. Soil organic carbon pools in the northern circumpolar permafrost region. Global Biogeochemical Cycles 23(2).

Taylor, C. M., R. A. M. de Jeu, F. Guichard, P. P. Harris and W. A. Dorigo. 2012. Afternoon rain more likely over drier soils. Nature 489(7416):423-426.

Taylor, J. E., J. Hardner and M. Stewart. 2008. Ecotourism and economic growth in the Galapagos: An island economy-wide analysis. Environmental and Developmental Economics 14:139-162.

Taylor, K. C., G. W. Lamorey, G. A. Doyle, R. B. Alley, P. M. Grootes, P. A. Mayewski, J. W. C. White and L. K. Barlow. 1993. The flickering switch of late Pleistocene climate change. Nature 361(6411):432-436.

Tebaldi, C., B. H. Strauss and C. E. Zervas. 2012. Modelling sea level rise impacts on storm surges along US coasts. Environmental Research Letters 7(1).

Tercek, M. R. and J. S. Adams. 2013. Nature's Fortune: How Business and Society Thrive by Investing in Nature. New York: Basic Books.

Thompson, D. W. J., S. Solomon, P. J. Kushner, M. H. England, K. M. Grise and D. J. Karoly. 2011. Signatures of the Antarctic ozone hole in Southern Hemisphere surface climate change. Nature Geoscience 4(11):741-749.

Thorndike, A. S. 1992. A toy model linking atmospheric thermal-radiation and sea ice growth. Journal of Geophysical Research-Oceans 97(C6):9401-9410.

Thuiller, W., O. Broennimann, G. Hughes, J. R. M. Alkemade, G. F. Midgley and F. Corsi. 2006. Vulnerability of African mammals to anthropogenic climate change under conservative land transformation assumptions. Global Change Biology 12(3):424-440.

Tietsche, S., D. Notz, J. H. Jungclaus and J. Marotzke. 2011. Recovery mechanisms of Arctic summer sea ice. Geophysical Research Letters 38(2).

Tietsche, S., D. Notz, J. H. Jungclaus and J. Marotzke. 2013. Predictability of large interannual Arctic sea-ice anomalies. Climate Dynamics 41(9-10):2511-2526.

Tilman, D., C. Balzer, J. Hill and B. L. Befort. 2011. Global food demand and the sustainable intensification of agriculture. Proceedings of the National Academy of Sciences of the United States of America 108(50):20260-20264.

Tindall, J., R. Flecker, P. Valdes, D. N. Schmidt, P. Markwick and J. Harris. 2010. Modelling the oxygen isotope distribution of ancient seawater using a coupled ocean-atmosphere GCM: Implications for reconstructing early Eocene climate. Earth and Planetary Science Letters 292(3-4):265-273.

Tol, R. S. J. 2003. Is the uncertainty about climate change too large for expected cost-benefit analysis? Climatic Change 56(3):265-289.

Townsend, K. E. B., D. T. Rasmussen, P. C. Murphey and E. Evanoff. 2010. Middle Eocene habitat shifts in the North American western interior: A case study. Palaeogeography, Palaeoclimatology, Palaeoecology 297:144-158.

Trenberth, K. E., P. D. Jones, P. Ambenje, R. Bojariu, D. Easterling, A. K. Tank, D. Parker, F. Rahimzadeh, J. A. Renwick, M. Rusticucci, B. Soden and P. Zhai. 2007. Observations: Surface and Atmospheric Climate Change. In Climate Change 2007: The Physical Science Basis. Contribution of Working Group I to the Fourth Assessment Report of the Intergovernmental Panel on Climate Change. Solomon, S., D. Qin, M. Manning, Z. Chen, M. Marquis, K. B. Averyt, M. Tignor and H. L. Miller, eds. Cambridge, UK and New York: Cambridge University Press.

Turner, J., T. J. Bracegirdle, T. Phillips, G. J. Marshall and J. S. Hosking. 2013. An initial assessment of Antarctic sea ice extent in the CMIP5 Models. Journal of Climate 26(5):1473-1484.

U.S. CLIVAR Project Office. 2011. Third Annual Progress Report for a JSOST Near-Term Priority Assessing Meridional Overturning Circulation Variability: Implications for Rapid Climate Change, Report 2011-1. Washington, DC.

USAID (U.S. Agency for International Development). 2013. Kenya, Environment. USAID Kenya: http://kenya.usaid.gov/programs/environment.

USCCSP (U.S. Climate Change Science Program). 2008. Synthesis and Assessment Product 3.4: Abrupt Climate Change. Washington, DC: U.S. Department of the Interior.

USGCRP (U.S. Global Change Research Program). 2009. Global Climate Change Impacts in the United States. New York: Cambridge University Press.

van Vuuren, D. P., L. B. Bayer, C. Chuwah, L. Ganzeveld, W. Hazeleger, B. van den Hurk, T. van Noije, B. O'Neill and B. J. Strengers. 2012. A comprehensive view on climate change: coupling of earth system and integrated assessment models. Environmental Research Letters 7(2).

Vavrus, S. J., M. M. Holland, A. Jahn, D. A. Bailey and B. A. Blazey. 2012. Twenty-first-century Arctic climate change in CCSM4. Journal of Climate 25(8):2696-2710.

Vecchi, G. A., B. J. Soden, A. T. Wittenberg, I. M. Held, A. Leetmaa and M. J. Harrison. 2006. Weakening of tropical Pacific atmospheric circulation due to anthropogenic forcing. Nature 441(7089):73-76.

Vecchi, G. A. and A. T. Wittenberg. 2010. El Niño and our future climate: Where do we stand? Wiley Interdisciplinary Reviews-Climate Change 1(2):260-270.

Veraart, A. J., E. J. Faassen, V. Dakos, E. H. van Nes, M. Lurling and M. Scheffer. 2012. Recovery rates reflect distance to a tipping point in a living system. Nature 481(7381):357-U137.

Vitousek, P. M., P. R. Ehrlich, A. H. Ehrlich and P. A. Matson. 1986. Human appropriation of the products of photosynthesis. BioScience 36:368-373.

Vitousek, P. M., H. A. Mooney, J. Lubchenco and J. M. Melillo. 1997. Human domination of Earth's ecosystems. Science 277:494-499.

Vorosmarty, C. J., P. Green, J. Salisbury and R. B. Lammers. 2000. Global water resources: Vulnerability from climate change and population growth. Science 289(5477):284-288.

Wada, Y., L. P. H. v. Beek, F. C. S. Weiland, B. F. Chao, Y. H. Wu and M. F. P. Bierkens, Past and future contributions of global groundwater depletion to sea-level rise. 2012. Past and future contributions of global groundwater depletion to sea-level rise. Geophysical Research Letters 39(L09402).

Walker, R. T., B. R. Parizek, R. B. Alley, S. Anandakrishnan, K. L. Riverman and K. Christianson. 2013. Ice-shelf tidal flexure and subglacial pressure variations. Earth and Planetary Science Letters 361:422-428.

Wallace, J. M., C. Deser, B. V. Smoliak and A. S. Phillips. 2013. Attribution of climate change in the presence of internal variability. In Climate Change: Multidecadal and Beyond. Chang, C. P., M. Ghil, M. Latif and J. M. Wallace, eds. Hackensack, NJ: World Scientific Publishing Company.

Walter, K. M., L. C. Smith and F. S. Chapin. 2007. Methane bubbling from northern lakes: present and future contributions to the global methane budget. Philosophical Transactions of the Royal Society A—Mathematical Physical and Engineering Sciences 365(1856):1657-1676.

Walther, G.-R., E. Post, P. Convey, A. Menzel, C. Parmesan, T. J. C. Beebee, J.-M. Fromentin, O. Hoegh-Guldberg and F. Bairlein. 2002. Ecological responses to recent climate change. Nature(6879):389-395.

Wang, G. L. and E. A. B. Eltahir. 2000. Biosphere-atmosphere interactions over West Africa. II: Multiple climate equilibria. Quarterly Journal of the Royal Meteorological Society 126(565):1261-1280.

Wang, M. Y. and J. E. Overland. 2009. A sea ice free summer Arctic within 30 years? Geophysical Research Letters 36(7).

Wang, M. Y. and J. E. Overland. 2012. A sea ice free summer Arctic within 30 years: An update from CMIP5 models. Geophysical Research Letters 39(18).

Wang, R., J. A. Dearing, P. G. Langdon, E. Zhang, X. Yang, V. Dakos and M. Scheffer. 2012a. Flickering gives early warning signals of a critical transition to a eutrophic lake state. Nature 492:419-422.

Wang, S. G., E. P. Gerber and L. M. Polvani. 2012b. Abrupt circulation responses to tropical upper-tropospheric warming in a relatively simple stratosphere-resolving AGCM. Journal of Climate 25(12):4097-4115.

Wang, Y. J., H. Cheng, R. L. Edwards, X. G. Kong, X. H. Shao, S. T. Chen, J. Y. Wu, X. Y. Jiang, X. F. Wang and Z. S. An. 2008. Millennial- and orbital-scale changes in the East Asian monsoon over the past 224,000 years. Nature 451(7182):1090-1093.

Warren, R., S. D. Santos, N. W. Arnell, M. Bane, T. Barker, C. Barton, R. Ford, H. M. Fussel, R. K. S. Hankin, R. Klein, C. Linstead, J. Kohler, T. D. Mitchell, T. J. Osborn, H. Pan, S. C. B. Raper, G. Riley, H. J. Schellnhuber, S. Winne and D. Anderson. 2008. Development and illustrative outputs of the Community Integrated Assessment System (CIAS), a multi-institutional modular integrated assessment approach for modelling climate change. Environmental Modelling & Software 23(5):592-610.

Weaver, A. J., J. Sedlacek, M. Eby, K. Alexander, E. Crespin, T. Fichefet, G. Philippon-Berthier, F. Joos, M. Kawamiya, K. Matsumoto, M. Steinacher, K. Tachiiri, K. Tokos, M. Yoshimori and K. Zickfeld. 2012. Stability of the Atlantic meridional overturning circulation: A model intercomparison. Geophysical Research Letters 39(20).

Weber, S. L., S. S. Drijfhout, A. Abe-Ouchi, M. Crucifix, M. Eby, A. Ganopolski, S. Murakami, B. Otto-Bliesner and W. R. Peltier. 2007. The modern and glacial overturning circulation in the Atlantic Ocean in PMIP coupled model simulations. Climate of the Past 3(1):51-64.

Weitzman, M. L. 2009. On modeling and interpreting the economics of catastrophic climate change. Review of Economics and Statistics 91(1):1-19.

Weitzman, M. L. 2011. Fat-tailed uncertainty in the economics of catastrophic climate change. Review of Environmental Economics and Policy 5(2):275-292.

Westra, S., L. V. Alexander and F. W. Zwiers. 2013. Global increasing trends in annual maximum daily precipitation. Journal of Climate 26:3904-3918.

Wettstein, J. and C. Deser. 2013. Internal variability in projections of twenty-first century Arctic sea ice loss: Role of the large-scale atmospheric circulation. Journal of Climate, in press.

Whiteman, G., C. Hope and P. Wadhams. 2013. Vast costs of Arctic change. Nature 499(7459):401-403.

WHO (World Health Organization). 2000. Climate Change and Human Health: Impact and adaptation. Geneva, Switzerland: World Health Organization.

WHO/WMO. 2012. Atlas of Health and Climate. Geneva, Switzerland: World Health Organization.

Wieczorek, S., P. Ashwin, C. M. Luke and P. M. Cox. 2011. Excitability in ramped systems: The compost-bomb instability. Proceedings of the Royal Society A-Mathematical Physical and Engineering Sciences 467(2129):1243-1269.

Williams, J. W. and S. T. Jackson. 2007. Novel climates, no-analog communities, and ecological surprises. Frontiers in Ecology and the Environment 5:475-482.

Williams, J. W., S. T. Jackson and J. E. Kutzbach. 2007. Projected distributions of novel and disappearing climates by 2100 AD. Proceedings of the National Academy of Sciences of the United States of America104:5738-5742.

Willis, J. K. 2010. Can in situ floats and satellite altimeters detect long-term changes in Atlantic Ocean overturning? Geophysical Research Letters 37(6).

Winton, M. 2006. Does the Arctic sea ice have a tipping point? Geophysical Research Letters 33(23).

Winton, M. 2008. Sea ice-albedo feedback and nonlinear Arctic climate change. In Arctic Sea Ice Decline: Observations, Projections, Mechanisms, and Implications, Geophys. Monogr. Ser., 180. DeWeaver, E. T., C. M. Bitz and L.-B. Tremblay, eds. Washington, DC: American Geophysical Union.

WMO/UNEP (World Meterological Organization/ United Nations Environment Programme). 2010. Scientific Assessment of Ozone Depletion: 2010. World Meterological Organization Global Ozone Research and Monitoring Project—Report No. 52. Geneva, Switzerland: World Meterological Organization.

Wood, R. A., A. B. Keen, J. F. B. Mitchell and J. M. Gregory. 1999. Changing spatial structure of the thermohaline circulation in response to atmospheric CO_2 forcing in a climate model. Nature 399:572-575.

Woods and Poole Economics Inc. 2011. Metropolitan Area Projections to 2040: Complete Economic and Demographic Data for Every Metropolitan Statistical Area (MSA) and Micropolitan Statistical Area in the United States. Washington, DC: Woods and Poole Economics Inc.

Wootton, J. T. and C. A. Pfister. 2012. Carbon system measurements and potential climatic drivers at a site of rapidly declining ocean pH. PLoS ONE 7(12).

World Water Council. 2000. The Use of Water Today. In World Water Vision: Making Water Everybody's Business. Cosgrove, W. J. and F. R. Rijsberman, eds. London: Earthscan Publications Ltd.

WRI (World Resources Institute). 2005. Millennium Ecosystem Assessment, Ecosystems and Human Well-being: Biodiversity Synthesis. Washington, DC: World Resources Institute.

Yang, M. X., F. E. Nelson, N. I. Shiklomanov, D. L. Guo and G. N. Wan. 2010. Permafrost degradation and its environmental effects on the Tibetan Plateau: A review of recent research. Earth-Science Reviews 103(1-2):31-44.

Yin, J. H. 2005. A consistent poleward shift of the storm tracks in simulations of 21st century climate. Geophysical Research Letters 32(18).

Yin, L., R. Fu, E. Shevliakova and R. E. Dickinson. 2012. How Well Can CMIP5 Simulate Rainfall Seasonal and Interannual Variability over Amazonian and South American Monsoon Regions and Their Controlling Processes? Climate Dynamics. doi: 10.1007/s00382-012-1582-y.

Yoshikawa, K., W. R. Bolton, V. E. Romanovsky, M. Fukuda and L. D. Hinzman. 2002. Impacts of wildfire on the permafrost in the boreal forests of Interior Alaska. Journal of Geophysical Research-Atmospheres 108(D1).

Youngblut, C. 2010. Climate Change Effects: Issues for International and US National Security. Alexandria, VA: Institute for Defense Analyses.

Zelazowski, P., Y. Malhi, C. Huntingford, S. Sitch and J. B. Fisher. 2011. Changes in the potential distribution of humid tropical forests on a warmer planet. Philosophical Transactions of the Royal Society S-Mathematical Physical and Engineering Sciences 369(1934):137-160.

Zeng, N., R. E. Dickinson and X. B. Zeng. 1996. Climatic impact of Amazon deforestation—A mechanistic model study. Journal of Climate 9(4):859-883.

Zeng, N. and J. D. Neelin. 2000. The role of vegetation-climate interaction and interannual variability in shaping the African savanna. Journal of Climate 13(15):2665-2670.

Zhang, C. D. 2005. Madden-Julian oscillation. Reviews of Geophysics 43(2).

Zhang, D. D., H. F. Lee, C. Wang, B. S. Li, Q. Pei, J. Zhang and Y. L. An. 2011. The causality analysis of climate change and large-scale human crisis. Proceedings of the National Academy of Sciences of the United States of America 108(42):17296-17301.

Zhang, Y., R. Fu, H. B. Yu, Y. Qian, R. Dickinson, M. A. F. S. Dias, P. L. D. Dias and K. Fernandes. 2009. Impact of biomass burning aerosol on the monsoon circulation transition over Amazonia. Geophysical Research Letters 36(10).

Zhang, Y., J. M. Wallace and D. S. Battisti. 1997. ENSO-like interdecadal variability: 1900-93. Journal of Climate 10(5):1004-1020.

Zhao, M., I. M. Held, S. J. Lin and G. A. Vecchi. 2009. Simulations of Global Hurricane Climatology, Interannual Variability, and Response to Global Warming Using a 50-km Resolution GCM. Journal of Climate 22(24):6653-6678.

Zhu, K., C. W. Woodall and J. S. Clark. 2012. Failure to migrate: lack of tree range expansion in response to climate change. Global Change Biology 18(3):1042-1052.

Zimov, S. A., S. P. Davydov, G. M. Zimova, A. I. Davydova, E. A. G. Schuur, K. Dutta and F. S. Chapin. 2006a. Permafrost carbon: Stock and decomposability of a globally significant carbon pool. Geophysical Research Letters 33(20).

Zimov, S. A., E. A. G. Schuur and F. S. Chapin. 2006b. Permafrost and the global carbon budget. Science 312(5780):1612-1613.

Biographical Sketches of Committee Members

Dr. James W.C. White (*Chair*) is a professor of Geological Sciences and of Environmental Studies at the University of Colorado at Boulder, where he is also a Fellow at the Institute of Arctic and Alpine Research (INSTAAR) and past director of the Environmental Studies Program. His research interests at the Light Stable Isotope Laboratory include global scale climate and environmental dynamics, carbon dioxide concentrations and climate from stable hydrogen isotopes, peats, and other organics, climate from deuterium excess and hydrogen isotopes in ice cores, isotopes in general circulation models, and modern carbon cycle dynamics via isotopes of carbon dioxide and methane. Dr. White has served on the Global Change Subcommittee, Planning Group 2, of SCAR from 1993 to 1996 and as a member of the US Ice Core Working Group from 1989 to 1992, after which he was the Chair from 1992 to 1996. He has served on the Polar Research Board of the National Research Council since May 2005. He was a member of the US Global Change Research Program's Synthesis and Assessment Product 1.2, Past Climate Variability and Change in the Arctic and at High Latitudes, from 2008-2009. Dr. White received his doctorate in Geological Sciences in 1983 from Columbia University. He is nominated to the committee for his knowledge of ice-sheet geochemistry and in particular of the materials that may enter subglacial environments from the overlying ice.

Dr. Richard B. Alley (NAS) is the Evan Pugh Professor of the Department of Geosciences and EMS Environment Institute at Pennsylvania State University. Dr. Alley studies past climate change by analyzing ice cores from Greenland and Antarctica. He has helped demonstrate that exceptionally large climate changes have occurred in as little as a single year. His work on deformation of subglacial tills has helped lead to new insights to ice-sheet stability and the interpretation of glacial deposits, and his ongoing work on ice-flow modeling may help lead to predictions of future sea-level change. Related interests include metamorphic textures of ice, transformation of snow to ice, microwave remote sensing of ice, origins of ice stratification, controls on snowfall, monitoring of past storm tracks. Along with his many teaching accomplishments, Dr. Alley has authored many publications, chaired the National Academy of Sciences' 2002 panel on abrupt climate change, has been involved with advisory groups to improve national and international research, and has been active with media outreach translate research findings to a broad audience with appearances on television, radio and print outlets. Dr. Alley received his Ph.D. in Geology at the University of Wisconsin, Madison.

Dr. David Archer is a professor in the Department of Geophysical Sciences at the University of Chicago. He earned his Ph.D. in oceanography from the University of Washington in 1990. He has worked on a wide range of topics pertaining to the global carbon cycle and its relation to global climate, with special focus on ocean sedimentary processes such as $CaCO_3$ dissolution and methane hydrate formation, and their impact on the evolution of atmospheric CO_2. He previously served on the NRC Committee on "Ocean Acidification: A National Strategy to Meet the Challenges of a Changing Ocean."

Dr. Anthony D. Barnosky is a Professor in the Department of Integrative Biology at The University of California, Berkeley. His research interests include understanding how global change influences biodiversity, evolution, and extinction, particularly for mammals. He received his Ph.D. in Geological Sciences from the University of Washington in 1983.

Dr. Jonathan Foley is the director of the Institute on the Environment (IonE) at the University of Minnesota, where he is a professor and McKnight Presidential Chair in the Department of Ecology, Evolution and Behavior. Dr. Foley's work focuses on the sustainability of our civilization and the global environment. He and his students have contributed to our understanding of global food security, global patterns of land use, the behavior of the planet's climate, ecosystems and water cycle, and the sustainability of the biosphere. This work has led him to be a regular advisor to large corporations, NGOs and governments around the world.

Dr. Rong Fu is a professor in the Department of Geological Sciences of the Jackson School of Geosciences at The University of Texas, Austin. Dr. Fu's research aims at understanding the dynamic and physical processes of the atmosphere's hydrological and energy cycles and their links to the terrestrial ecosystem and ocean surface conditions in the tropics using a variety of observations, especially those from satellite remote sensing. She has served on national and international panels such as the review panels for NASA Carbon Cycle Science Program, Cloud and Aerosol Program, NOAA Cooperative Institute for Climate Science at Princeton University, NSF Drought in Coupled Models Project, and panels for the U.S. and International Climate Variability and Predictability projects. She has also served on NRC Committee on Challenges and Opportunities in Earth Surface Processes. She received her B.S. degree in geophysics from Peking University in 1984, and her Ph.D. in atmospheric sciences from Columbia University in 1991. She then worked as a post-doctoral research associate at the Department of Atmospheric Sciences in the University of California, Los Angeles, and as a visiting scientist fellow at the Geophysical Fluid Dynamic Laboratory at Princeton University. She was previously a faculty member in the School of Earth and Atmospheric Sciences at Georgia Institute of Technology and in the Department of Atmospheric Sciences at the University of Arizona.

Dr. Marika Holland is a an Ice Specialist in the Oceanography section of the Climate and Global Dynamics division at the National Center for Atmospheric Research (NCAR). She received her Ph.D. in 1997 from the Program in Atmosphere and Ocean Sciences at the University of Colorado in the area of sea ice modeling for climate applications. Her training continued with a postdoctoral fellowship at the University of Victoria in British Columbia studying the influence of sea ice variability and change on the global ocean circulation and climate. In 1999, Dr. Holland moved to the NCAR in Boulder, Colorado, as a Postdoctoral Fellow and joined the scientific staff in 2000. Her research interests include polar climate variability and future change, including the role of ice-ocean-atmosphere interactions and feedbacks. She has extensive experience using coupled climate models to study these issues and has been active in the development of improved sea ice models for climate simulations. She is currently serving as Chief Scientist for the Community Earth System Modeling Project.

Dr. Susan Lozier is a physical oceanographer with interests in large-scale ocean circulation. Upon completion of her Ph.D. at the University of Washington, she was a post-doctoral scholar at Woods Hole Oceanographic Institution. She has been a member of the Duke faculty since 1992, where she was named a distinguished professor in 2012. Professor Lozier was the recipient of an NSF Early Career Award in 1996, was awarded a Bass Chair for Excellence in Research and Teaching in 2000, received a Duke University Award for Excellence in Mentoring in 2007 and was named an AMS Fellow in 2008.

Dr. Johanna Schmitt (NAS) is a professor in the Ecology and Evolutionary Biology Department at Brown University. Dr. Schmitt's research focus is adaptive evolution of developmental, physiological, and life history traits in natural plant populations. Her lab uses quantitative genetics, QTL mapping, and association studies of candidate loci to examine the genetic basis of natural variation in ecologically important traits. By experimentally manipulating environments, phenotypes, and genotypes in the field, they measure natural selection on these traits and the loci underlying them. Another major research objective is to elucidate the genetic and ecological mechanisms of adaptation to seasonal and geographic variation in climate. Dr. Schmitt received her Ph.D. in Biological Sciences from Stanford University in 1981.

Dr. Laurence C. Smith is Professor and Vice Chair of Geography and Professor of Earth & Space Sciences at the University of California, Los Angeles. His research interests include topics of northern hydrology, climate change, carbon cycles, and satellite remote sensing. In 2007 his work appeared prominently in the Fourth Assessment Report of the United Nations' Intergovernmental Panel on Climate Change (IPCC). In 2006-2007 he was named a Guggenheim Fellow by the John S. Guggenheim Foundation. Dr. Smith received his Ph.D. in Earth and Atmospheric Sciences in 1996 from Cornell University.

Dr. George Sugihara is a professor and department chair at SIO at the University of California, San Diego. He earned his Ph.D. in Mathematical Biology from Princeton. His diverse research interests include complexity theory, nonlinear dynamics, food web structure, species abundance patterns, conservation biology, biological control, empirical climate modeling, fisheries forecasting, and the design and implementation of derivative markets for fisheries. One of his most interdisciplinary contributions involves the work he developed with Robert May concerning methods for forecasting nonlinear and chaotic systems. This took him into the arena of investment banking, where he took a five-year leave from SIO to become Managing Director for Deutsche Bank. There he made a successful application of these theoretical methods to forecast erratic market behavior. Most of Dr. Sugihara's early work was motivated exclusively by pure science and the later work more by pragmatic utility and environmental concerns. Nearly all of it is based on extracting information from observational data (turning data into information). His initial work on fisheries as complex, chaotic systems led to work on financial networks and prediction of chaotic systems. Dr. Sugihara serves on the Board on Mathematical Sciences and their Applications at the NRC and also served on the Planning Committee for a Workshop on Technical Capabilities Required for Regulation of Systemic Risk.

Dr. David Thompson is a Professor in the Department of Atmospheric Science at Colorado State University. Dr. Thompson's research focuses on improving our understanding of global climate variability using observational data. His interests include large-scale atmospheric dynamics, the interpretation of observed climate change, stratosphere/troposphere coupling, ocean/atmosphere interaction, decadal climate variability, and the climate impacts of large-scale atmospheric phenomena. Dr. Thompson received his Ph.D. in Atmospheric Science in 2000 from the University of Washington.

Dr. Andrew Weaver is a Lansdowne Professor and Canada Research Chair at the University of Victoria, British Columbia. Dr. Weaver's research focuses upon understanding processes and feedbacks operating within the climate system on a range of timescales. In particular, he is interested in exploring the role of the oceans in past, present and future climate change using his locally developed Earth Model of Intermediate

Complexity (the UVic ESCM). Dr. Weaver received his Ph.D. in 1987 from the University of British Columbia.

Dr. Steven C. Wofsy (NAS) was born in New York City in 1946 and is currently Abbott Lawrence Rotch Professor of Atmospheric and Environmental Chemistry at Harvard University, Division of Engineering and Applied Science and Department of Earth and Planetary Sciences. He studied chemical physics at University of Chicago (B.S., 1966) and Harvard (Ph.D., 1971), shifting to atmospheric chemistry in 1971. His work has focused on changes in the composition of the stratosphere and troposphere, at first in theory and modeling and later in field and laboratory studies. His current research emphasizes the effects of terrestrial ecosystems on the global carbon cycle, aircraft measurements of greenhouse gases in the atmosphere, the impacts of climate change and land use on ecosystems and atmospheric composition. Several projects focus on quantitative measurements of ecosystem carbon fluxes, for time scales spanning instantaneous to decadal and spatial scales from meters to thousands of kilometers, combining physical, chemical and biological methods. He is a member of the National Academy of Sciences. His awards include AGU's McIlwane prize and NASA's Distinguished Public Service Medal.